MW00838117

Roughness Characteristics of New Zealand Rivers

by D M Hicks and P D Mason

A handbook for assigning hydraulic roughness coefficients to river reaches by the "visual comparison" approach

National Institute of Water and Atmospheric Research Ltd
September 1998
Water Resources Publications, LLC

Waiau Water Race at Lateral 2 (pages 110-113)
$Q = 2.12 \ m^3/s, \ n = 0.027$

ABSTRACT

Physical and hydraulic characteristics are presented for 78 New Zealand river and canal reaches which may be used as reference reaches for estimating roughness coefficients in similar channels. These reaches cover a wide range of mean flow (ranging from 0.1 to 353 m³/s), slope (ranging from 0.00001 to 0.042), and bed material (ranging from silt, through sand, gravel, and boulders to bedrock). The information given for each reach includes colour photographs, cross-section and planform plots, bed and bank descriptions, bed surface material size gradings, plus tables and plots showing how the Manning and Chezy roughness coefficients vary with flow magnitude.

ACKNOWLEDGEMENTS

The information presented here was collected by field staff of the DSIR Water Resources Survey, located in Whangarei, Auckland, Hamilton, Wanganui, Rotorua, Turangi, Gisborne, Havelock North, Wellington, Nelson, Greymouth, Christchurch, Lake Tekapo, Alexandra, and Dunedin. Gavin Sullivan assisted the authors with data preparation and plots. The project was conceived and managed by M.P. Mosley.

Roughness Characteristics of New Zealand Rivers

Reprint of ISBN 0-477-02608-7 by

National Institute of Water and Atmospheric Research Ltd
Christchurch, New Zealand
September 1998

First printed and published by

Water Resources Survey
DSIR Marine and Freshwater
Wellington, New Zealand
June 1991

Distributed by:

Water Resources Publications, LLC
7200 E. Dry Creek Rd., Suite E - 104
Englewood, Co 80112, USA
Telephone (303) 741 9071

CONTENTS

Page

Abstract iii

Acknowledgements iv

Symbols vi

Introduction 1

Reach selection 2

Field measurements 3

Calculation of roughness coefficients and other parameters 4

Errors and uncertainties 8

Presentation of information 9

How to use this book 11

Quick method for selecting reference reaches 12

Site information 14

References 326

Site index 327

SYMBOLS

A	Area of wetted channel cross-section
ΔA	Percentage expansion in area from upstream to downstream cross-section
α	Velocity head coefficient
C	Chezy's roughness coefficient
d	Diameter of bed surface material
d_{50}	Median diameter of bed surface material
g	Gravitational acceleration
h	Water surface elevation
h_f	Head loss due to boundary friction
h_v	Velocity head
Δh	Upstream water surface elevation minus downstream water surface elevation
Δh_v	Upstream velocity head minus downstream velocity head
k	A coefficient for defining energy losses due to diverging or converging flow
L	Length of reach
m	Number of cross-sections along a reach
n	Manning's roughness coefficient
Q	Water discharge
R	Hydraulic radius of channel cross-section
S_f	Friction slope
S_w	Water surface slope
V	Mean velocity
X	A factor equal to $AR^{1/2}$
Z	A factor equal to $AR^{2/3}$

INTRODUCTION

The information presented here is the culmination of a three-year field programme in which roughness and other hydraulic parameters were measured at 78 reaches representing a broad range of New Zealand rivers. The aim of the programme was to provide a reference dataset for use in visually estimating roughness coefficients. This responded to a need for a reference set of reaches representative of New Zealand conditions - our own combination of channel size, gradient, bed material, and vegetation - that would also cater for variations in roughness with discharge.

The standard references for estimating roughness coefficients by the "visual comparison" approach have been Chow (1959) and Barnes (1967). Chow features only 24 reaches, most of which are in artificial channels. Barnes is more comprehensive, describing 50 natural reaches typical of many North American rivers. However, while there are close physical similarities between many North American and New Zealand rivers, there are often significant differences in bank vegetation. Also, the Barnes dataset rarely presents more than one measurement of Manning's *n* per site, whilst it is well known that Manning's *n* can vary considerably with discharge.

Values of roughness coefficient estimated using the "visual comparison" approach can be employed in their own right in hydraulic calculations; alternatively, and preferably, they can be used either to verify, or as base values for, estimates of roughness coefficients obtained by more quantitative methods. One quantitative approach involves empirically derived predictive equations, such as those of Bray (1979), Griffiths (1981), and Jarrett (1984). Griffiths' empirical equations were derived from 136 gaugings at 72 reaches in New Zealand gravel-bed rivers. Another approach, developed by Cowan (1956), involves first assigning a base-value roughness coefficient appropriate for a straight, uniform, smooth channel with the same bed material as the reach under investigation, then increasing this base-value by applying correction factors for the effects of bed surface irregularities, shape and size of the cross-section, obstructions, vegetation, and meandering. The base-value can be assigned by visual comparison, empirical equation, or both, provided that the reference data are from channels meeting the "straight and uniform" requirements. Chow (1959) and Arcement and Schneider (1989) provide base-values and correction factors for Manning's *n* for use with this approach.

1

The following sections describe the rationale for reach selection, the methods of calculating roughness coefficients and other properties of the study reaches, errors that enter these calculations, the manner of presentation of the information for each reach, and an outline of how this handbook could be used for estimating roughness, including a quick method for selecting a reference reach. The remainder and bulk of the document presents detailed information on each of the study reaches.

REACH SELECTION

The 78 study reaches were selected primarily to provide the broadest possible coverage of river size (as indexed by mean flow), channel gradient, and bed material. A second criterion was that reaches should - as far as possible - be typical "slope-area" reaches, conveying their flow uniformly through a single channel. Ideal characteristics of a slope-area reach are that (after Arnold et al., 1988):

(i)   it is straight

(ii) its length is at least five times its width

(iii) it has uniform cross-sections or is converging

(iv) its flow is contained without overflow

(v)   it has straight entrance and exit conditions, with no backwater effects.

Rarely, if ever, in the study reaches were all of these ideal conditions met. Thus the approach adopted is to present roughness information from "real" reaches that are as close as possible to uniform, and to include sufficient additional information so that readers can make their own assessment regarding uniformity and the extent to which a reference reach matches their prototype.

Where these first selection criteria were met, existing flow monitoring sites were used, otherwise new sites were installed.

As noted above, measurements were generally limited to flows confined within the banks of simple, single-thread channels. Readers are referred to Chow (1959) or Arcement and Schneider (1989) for guidelines on determining roughness coefficients for flows through compound sections involving floodplains and/or multi-thread channels.

FIELD MEASUREMENTS

All field measurements followed standard techniques which, unless specified otherwise below, are outlined in Arnold et al. (1988). Within each reach, cross-sections were established at reasonably uniform spacings and were surveyed once. The number of cross-sections per reach was generally from three to five, although several reaches had only two. Staff gauges were installed along one bank at each cross-section. Water-level was monitored continuously via float recorder or pressure transducer at at least one cross-section and combined with a flow rating to obtain a discharge record. In some reaches, pressure transducers or flood-crest recorders were installed at each cross-section.

Assessments of roughness coefficients were made over as wide a range of flows as possible. The number of assessments per reach averaged six to seven, but varied from two to 14. Most of the dataset was collected between March 1988 and December 1990, although at any given reach the data were usually collected over a shorter time frame, sometimes over a single high-flow event. For a few reaches, older data were included, provided that the reach was known to be stable and the measurements were of the required standard.

Each assessment involved determining the discharge through the reach and the water level at each cross-section. Discharge was determined either by current-meter gauging or from a reliable flow rating. The current-meters were usually cup-type but some were screw-type. The flow ratings were either based directly on current-meter gaugings or were theoretically based and verified/adjusted by gaugings. Water levels were generally obtained from staff gauges, pressure transducers, or flood-crest recorders. Less often, water levels were pegged at the assessment time then levelled later. For a few cases, water levels corresponding to flood crests were taken from silt lines, providing these could be precisely located on the banks. Where necessary along reaches of very gentle gradient, cross-section datums were coordinated by precise levelling using laser instruments. Hydraulic radius and wetted area were obtained from their respective hydraulic geometry relationships with stage, established from the cross-section surveys. At all but two reaches, visual observations and stable flow ratings indicated that the channel cross-sections remained stable throughout the study period. The two reaches where channel changes were observed are identified in the site information.

For gravel and boulder-bed channels where the bed was exposed or wadable at low flow, the size grading of the surface material was measured in the field by the Wolman method (Wolman, 1954). For sand-bed channels, representative samples were collected for sieve analysis in the laboratory. Where bed sampling was impractical, a visual characterisation of the bed surface material was made. These bed surface material assessments were made only once and it was assumed that they remained representative over the study period. The observed stability of cross-sections justifies this assumption.

Photographs of each reach were taken, generally during low flow conditions towards the end of the study period.

CALCULATION OF ROUGHNESS COEFFICIENTS AND OTHER PARAMETERS

Roughness Coefficients

Two roughness coefficients are determined: Manning's n and Chezy's C. The coefficient n is intended for use in the Manning Equation

$$Q = \frac{1}{n}AR^{2/3}S_f^{1/2} \tag{1}$$

where Q is the water discharge (in m^3/s), A is the wetted channel cross-sectional area (in m^2), R is the hydraulic radius (in m), and S_f is the friction slope. The friction slope is defined as

$$S_f = \frac{h_f}{L} = \frac{\Delta h + \Delta h_v - k(\Delta h_v)}{L} \tag{2}$$

where (as shown in Fig. 1) h_f is the head loss due to boundary friction along the reach, L is the reach length, Δh is the change in elevation of the water surface between the upstream and downstream cross-sections, Δh_v is the change in velocity head between the upstream and downstream cross-sections, and $k(\Delta h_v)$ approximates the energy loss due to acceleration or deceleration in a contracting or expanding reach. Following convention (e.g. Chow, 1959), k is assumed equal to zero for contracting reaches and 0.5 for expanding reaches[1]. The velocity head, h_v, at a cross-section is equal to $\alpha V^2/2g$, where g is the acceleration due to gravity, V is the mean velocity (equal to Q/A), and α is the velocity head coefficient, which indicates the uniformity of velocity across the channel. In this study, α is assumed equal to 1.0. This maintains consistency with previous workers who have presented roughness coefficient data (e.g. Chow, 1959; Barnes, 1963; Jarrett, 1984) and with the application of the slope-area method to channels with simple cross-sections.[2]

The coefficient C is for use in the Chezy Equation

$$Q = CAR^{1/2}S_f^{1/2} \tag{3}$$

Note that, from equations (1) and (3), the relationship between the Chezy and Manning roughness coefficients (neither of which are dimensionless) is

$$C = \frac{R^{1/6}}{n} \tag{4}$$

Following the method adopted by Barnes (1967) and Jarrett (1984), an expression for a representative value of Manning's n for a multi-section reach at a given discharge can be obtained by equating the friction head loss calculated from the friction slope in equation (1) with the friction head loss given by equation (2). That is,

from (1):

$$h_f = h_{f_{1.2}} + h_{f_{2.3}} + \ldots. h_{f_{(m-1).m}}$$

$$= n^2 Q^2 \left(\frac{L_{1.2}}{Z_1 Z_2} + \frac{L_{2.3}}{Z_2 Z_3} + \ldots. \frac{L_{(m-1).m}}{Z_{(m-1)}.Z_m} \right) \tag{5}$$

where m is the number of cross-sections (with the mth cross-section being furtherest upstream, as defined in Fig. 1), $Z = AR^{2/3}$, and a

---

[1] It is recognized that few measurements of k are available for natural rivers and these have shown that k can vary widely.

[2] Typically, α tends to have a value slightly greater than 1.0 for most of the channel conditions investigated in this study. However, a higher value of α, about 1.3, is likely in the mountain streams (Jarrett, 1984).

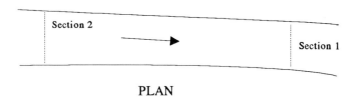

PLAN

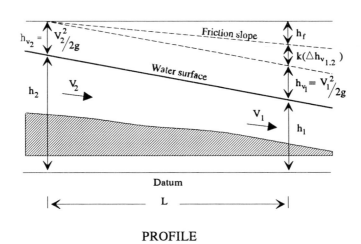

PROFILE

Figure 1. Definition sketch for a two-section reach.

representative value of Z for the reach between two adjacent cross-sections is $(Z_1 Z_2)^{1/2}$,

and from (2):

$$h_f = (h_m - h_1) + (h_{V_m} - h_{V_1})$$
$$- (k_{1.2} \Delta h_{V_{1.2}} + k_{2.3} \Delta h_{V_{2.3}} + \dots k_{(m-1).m} \Delta h_{V_{(m-1).m}}) \qquad (6)$$

thus from (5) and (6):

$$n = \frac{1}{Q} \left[\frac{(h_m - h_1) + (h_{V_m} - h_{V_1}) - (k_{1.2} \Delta h_{V_{1.2}} + \dots k_{(m-1).m} \Delta h_{V_{(m-1).m}})}{\dfrac{L_{1.2}}{Z_1 Z_2} + \dfrac{L_{2.3}}{Z_2 Z_3} + \dots \dfrac{L_{(m-1).m}}{Z_{(m-1)}.Z_m}} \right]^{1/2}$$

$$(7)$$

6

In similar fashion, a representative value of Chezy's C can be determined from

$$C = Q \left(\frac{\dfrac{L_{1.2}}{X_1 X_2} + \dfrac{L_{2.3}}{X_2 X_3} + \dots \dfrac{L_{(m-1).m}}{X_{(m-1)} X_m}}{(h_m - h_1) + (h_{V_m} - h_{V_1}) - (k_{1.2} \Delta h_{V_{1.2}} + \dots k_{(m-1).m} \Delta h_{V_{(m-1).m}})} \right)^{1/2}$$

(8)

where $X = AR^{1/2}$.

Equations (7) and (8) were used to calculate the values of n and C presented in this report.

Other Parameters

Reach-representative values of the other hydraulic parameters that are presented were determined as follows:

Friction slope, $\quad S_f$ - determined from equations (2) and (6)

Water surface slope, $\quad S_w = \dfrac{(h_m - h_1)}{L}$

Hydraulic radius, $\quad R = \dfrac{(R_1 + R_2 + \dots R_m)}{m}$

Wetted area, $\quad A = \dfrac{(A_1 + A_2 + \dots A_m)}{m}$

Mean velocity, $\quad V = \left(\dfrac{Q}{A_1} + \dfrac{Q}{A_2} + \dots \dfrac{Q}{A_m} \right) \dfrac{1}{m}$

Percentage expansion, $\quad \Delta A = 100 \dfrac{(A_1 - A_m)}{A_m}$

ΔA shows the extent to which the reach is either expanding (positive value) or contracting (negative value) between its upper and lower cross-sections.

7

ERRORS AND UNCERTAINTIES

The uncertainties in the calculated values of Manning's n and Chezy's C due to measurement error were determined by propagating through equations (7) and (8) the random and unknown systematic uncertainties associated with the measurements of the individual parameters. The uncertainty propagation followed the root-sum-square method (Herschy, 1985). Generally, the uncertainties used for the individual parameters (at the 95% confidence level) were in keeping with the standards maintained by the DSIR Water Resources Survey (Water Resources Survey, 1989). The uncertainty associated with a discharge value, either obtained by current meter gauging or from a rating[1], was taken as $\pm$ 8%. The uncertainty in a water level measurement was taken as $\pm$ 3 mm, except where the water level was fixed by pegging or from a silt-line, in which case an appropriately larger uncertainty was used[2]. The uncertainty in the datum level of the water level gauge or instrument was also taken as $\pm$ 3 mm, except where cross-sections were coordinated by precise levelling, in which case the uncertainty was reduced to $\pm$ 1 mm. The uncertainty in reach length was taken as $\pm$ 0.5%.

For most of the study reaches, the 95% confidence level uncertainties in the calculated roughness coefficients are dominated by the error in the discharge measurement and lie within the range $\pm$ 8% to $\pm$ 12%. Larger uncertainties typically occur at reaches with very gentle water surface slopes.

As well as being subject to measurement errors, the "reach representative" values of roughness coefficient calculated with equations (7) and (8) can also vary depending on the number and exact siting of cross-

---

[1] Unless the site had a flow-measurement structure with a reliable theoretical rating, rating curves were never extrapolated beyond a discharge 15% larger than the largest gauged discharge defining the rating.

[2] Errors in the values of wetted area and hydraulic radius used in equations (7) and (8) combine errors in the stage vs. area and stage vs. hydraulic radius relationships with errors in the water level measurement. Generally, where cross-sections remained stable over the study period, the former errors were assumed negligible, and so uncertainties in area and hydraulic radius became simple functions of the uncertainty of water level measurement. For the few sites where cross-sections did undergo significant change, the uncertainties associated with the stage vs. area and stage vs. hydraulic radius curves were estimated from changes observed in the flow rating.

sections in the reach. This is due to variations in roughness and channel geometry along natural river reaches and the constraint that these variations can only be sampled at a few sections. To assess this uncertainty, multiple determinations of Manning's n were made for the highest measured discharge at all reaches having five cross-sections, each time using one of the ten possible combinations of three cross-sections. The results showed that the 95% confidence level uncertainties on any single n determination based on three cross-sections were often large, averaging $\pm$ 21% for the 16 reaches evaluated; moreover, at lower discharges the uncertainties tended to increase as the relative uniformity of cross-section shape and water surface slope decreased. While not constituting a rigorous analysis of this source of error, the above serves to warn readers that the true uncertainties in the "reach-representative" values of n and C determined in this study will generally be larger than the uncertainties due to measurement error that are posted in the parameter tables.

PRESENTATION OF INFORMATION

Information

A four-page information set is presented for each reach covering general information on the site, the bed material, photographs, a tabulation of the hydraulic properties relating to each determination of roughness coefficient, plots of Manning's n against discharge and Chezy's C against hydraulic radius, a plan sketch of the reach, and cross-section plots.

The general site information includes: site name and number (as referenced in Walter, 1990); location, given as a grid reference for both the 1:50,000 scale NZMS260 (metric grid) and 1:63,360 scale NZMS1 (yard grid) map series; area of river catchment upstream of the site; period for which a flow record is available for the site; mean annual flood; and mean flow. Where possible, the mean annual flood values were taken from McKerchar and Pearson (1989); otherwise they were calculated as the average of annual peak flows over the period of flow record. Further information specific to this study includes: number of cross-sections and length of the reach used in the determination of the roughness coefficients; measured range in Manning's n values; and a description of the channel bed and banks.

Where a size analysis of the bed surface material was made, a plot is given showing the cumulative percentage by weight finer than a given grainsize and also a tabulation of the grainsizes at commonly used percentiles. For sand-bed channels, the percentage finer by weight data were obtained directly from sieving results. For gravel and boulder-bed channels, the percentage finer by weight was assumed equal to the percentage finer by count (after Kellerhals and Bray, 1971, and Church et al., 1987).

The photographs were taken from a variety of perspectives, depending on the field of view and accessibility afforded by bank vegetation. The camera positions and view directions are shown in the plan sketch.

The tabulation of hydraulic parameters includes the roughness coefficients, n and C, and their uncertainty due to measurement error (at the 95% confidence level), along with the associated values of water discharge, water surface slope, friction slope, percentage expansion of the reach between the upper and lower cross-sections, plus reach-representative values of hydraulic radius, cross-section area, and mean velocity. The latter three values are included to give an appreciation of the wetted channel's size and shape and to permit calculation of other parameters (e.g. Froude Number) as desired. The percentage expansion or contraction, plus a comparison of the water surface and friction slopes, can be used to assess the uniformity of flow through the reach.[1] Unless indicated otherwise, the water discharge values presented in the parameter tables were obtained by current meter gauging.

Changes in roughness with flow are demonstrated by the plots of Manning's n vs. discharge and of Chezy's C vs. hydraulic radius (generally in the form of a relative depth parameter, R/d_{50}, where d_{50} is the median bed surface particle size). For the n vs. Q plot the ordinate is plotted on a logarithmic scale purely to spread the data values, while on the C vs. R/d_{50} plot a logarithmic scale for R/d_{50} is consistent with theoretical and empirical relationships (e.g. Keulegan, 1938; Bray, 1979; Griffiths, 1981).

The plan sketch, not to scale, is presented to show the location of the photographs and cross-sections, the sinuosity, and the nature of the reach entrance and exit.

---

[1] Note that the percentage expansion is intended simply as an indicator of whether the reach is diverging or converging overall. It may not always be indicative of total expansion losses, as, for example, in reaches that expand or contract in their interior but have no substantial difference between their end sections. Thus readers should also take account of the cross-section and planform plots.

10

The three cross-section plots are of the upstream and downstream end sections, plus a central section. The shading shows the wetted area at the highest discharge used for a roughness assessment (i.e., the highest discharge value listed in the table of hydraulic parameters), while the white line shows the water level at the lowest discharge used.

Order of Presentation

The reaches are presented principally in order of increasing value of Manning's n at mean discharge. The broad range in n observed at many sites required that a single representative n value be indexed to some flow statistic; the mean flow was used here for no better reason than that both the mean flow and n at the mean flow (as estimated from the n vs. Q plot) were easily determined at most sites. These reach-representative values of Manning's n are shown in the page headers. In the few cases where the measurements were insufficient to allow n to be established at the mean flow, the representative n was taken simply as the average of the measured values, and the n value shown in the header is tagged "estimated". Where several reaches have the same representative n value, they are presented in order of increasing size of bed material.

HOW TO USE THIS BOOK

It is envisaged that this handbook will be used mainly to aid the assignment of roughness coefficients, for example, during the application of the slope-area method for estimating flood peak discharge. The suggested procedure for determining a roughness coefficient is as follows:[1]

1. Select a reference reach that is similar to the one being investigated by matching, as far as possible, channel size and shape, bed material, slope, bank vegetation, etc. A quick method for finding the page numbers of possible reference reaches is given overpage.

2. Once a reference reach has been chosen, a value of n or C can be assigned appropriate to the flow size. This may be done either by matching values of A, R, and/or S_w from the prototype reach with the data tabulated for the reference reach, or by interpolating from the plots of n vs. Q or C vs. R/d_{50}. Using the latter plot, given a value of R/d_{50}, C can be obtained directly and can, if required, be transformed to n via

---

[1] It is assumed that the reader will have available information on bed material and survey data on cross-sections and water surface slopes.

equation (4). Using the n vs. Q plot requires an iterative procedure if Q is not known. In this, an initial estimate of n obtained from the reference data tabulation is used with the prototype values of A, R, and S_w to calculate an initial estimate of Q via equation (1). This estimate of Q and the n vs. Q plot are then used to refine the estimate of n. The procedure should be repeated until the estimates of n converge.

3.   After a careful comparison of the characteristics of the prototype and reference reaches, the reader may wish to "fine-tune" the roughness coefficient to allow for differences in channel shape, sinuosity, vegetation, obstructions, etc., as outlined in Chow (1959) or Arcement and Schneider (1989).

QUICK METHOD FOR SELECTING REFERENCE REACHES

A quick route to finding the page numbers of suitable reference reaches is contained in Fig. 2. The information required is estimates of the mean discharge, water surface slope (ideally at the mean discharge but this can be approximated, e.g., by the slope obtained during a slope-area measurement), and bed material size at the reach being investigated. For example, for a fine gravel reach where the mean flow is 33 m³/s and a floodline slope was measured as 0.0008, then from Fig. 2 a likely reference reach can be found on page 34 or alternatively on page 90.

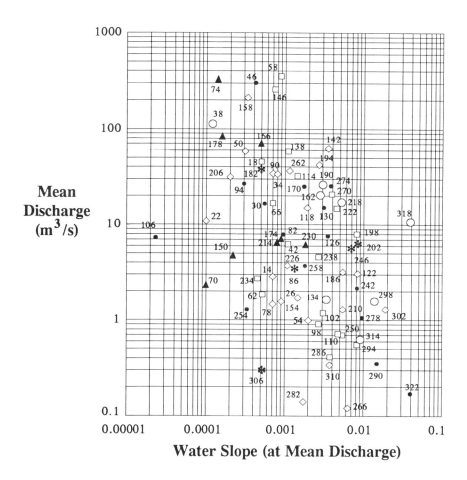

$$\text{Mean Discharge } (m^3/s)$$

Water Slope (at Mean Discharge)

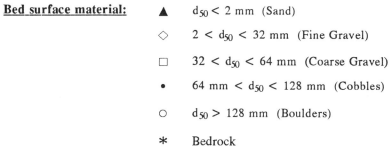

Bed surface material:

▲ $d_{50} < 2$ mm   (Sand)

◇ $2 < d_{50} < 32$ mm   (Fine Gravel)

□ $32 < d_{50} < 64$ mm   (Coarse Gravel)

• 64 mm $< d_{50} < 128$ mm   (Cobbles)

○ $d_{50} > 128$ mm   (Boulders)

✳   Bedrock

Figure 2. Diagram for finding page numbers of reference reaches.

$n = 0.016$

1643444: Poutu at Ford.

Map reference:-   T19:497328 (Metric); N112:256905 (Yard).
Catchment area:-   145 km^2.
Period of record:-   May 1983 - Present.
Mean annual flood:-   6 m^3/s.
Mean flow:-   2.86 m^3/s.

Surveyed reach:-
Cross-sections:-   2 along a 52 m reach.
Manning's n range:-   0.015-0.017
Channel description:-   Bed consists of an eroded lava flow. Banks are lined with grasses and Broom.

Bed Surface Material

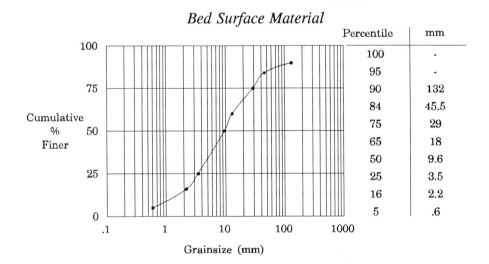

Percentile	mm
100	-
95	-
90	132
84	45.5
75	29
65	18
50	9.6
25	3.5
16	2.2
5	.6

View downstream of top cross-section.

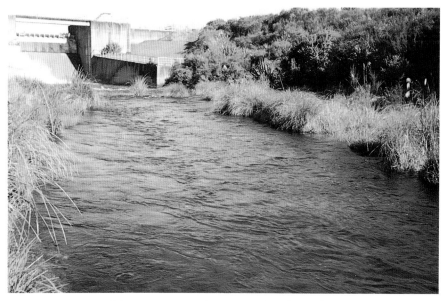

View upstream from middle of reach.

_n = 0.016_

Hydraulic Properties of Reach

Discharge	Water Surface Slope	Friction Slope	Area	Expansion	Hydraulic Radius	Mean Velocity	Manning n	Chezy C	Error
(m³/s)			(m²)	(%)	(m)	(m/s)			(%)
2.31	0.00067	0.00306	2.25	250	0.28	1.48	0.017	47.0	19
5.80	0.00103	0.00161	4.80	46	0.35	1.25	0.016	52.5	16
6.36	0.00077	0.00145	5.00	50	0.36	1.32	0.015	57.5	19

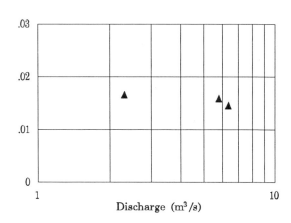

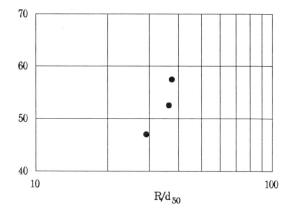

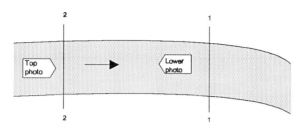

$n - 0.016$

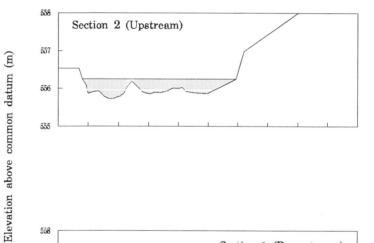

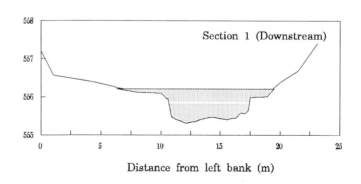

Plan (not to scale) and cross sections, Poutu at Ford.

n = 0.016 (est.)

23150: Ngaruroro at Chesterhope Bridge.

Map reference:-	V21:425715 (Metric); N134:289264 (Yard).
Catchment area:-	1994 km².
Period of record:-	November 1976 - December 1989.
Mean annual flood:-	949 m³/s.
Mean flow:-	45.6 m³/s.

Surveyed reach:-

Cross-sections:-	3 along a 742 m reach.
Manning's n range:-	0.015-0.019
Channel description:-	Bed consists of fine to medium gravel. Banks and berms are grassed.

Bed Surface Material

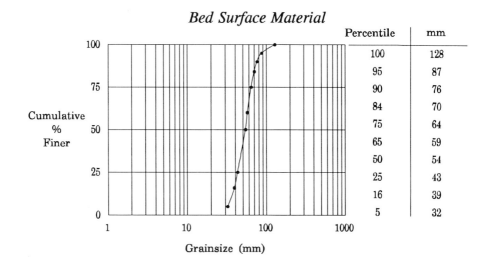

Percentile	mm
100	128
95	87
90	76
84	70
75	64
65	59
50	54
25	43
16	39
5	32

View downstream from bridge.

View upstream towards bridge.

n = 0.016 (est.)

Hydraulic Properties of Reach

Discharge (m³/s)	Water Surface Slope	Friction Slope	Area (m²)	Expansion (%)	Hydraulic Radius (m)	Mean Velocity (m/s)	Manning n	Chezy C	Error (%)
151*	0.00059	0.00062	77.1	13	1.24	1.97	0.015	71.6	10
165*	0.00063	0.00065	82.5	10	1.30	2.01	0.015	69.6	10
237*	0.00071	0.00074	127	13	1.55	1.86	0.019	55.3	9
268*	0.00075	0.00078	144	16	1.38	1.86	0.018	57.4	9
298*	0.00075	0.00079	161	18	1.33	1.85	0.018	57.6	9
341*	0.00073	0.00077	191	24	1.33	1.80	0.019	56.2	9
369*	0.00069	0.00069	215	30	1.34	1.74	0.019	55.9	9
563*	0.00064	0.00072	340	47	1.23	1.69	0.018	58.3	9

* Estimated from rating based on gaugings

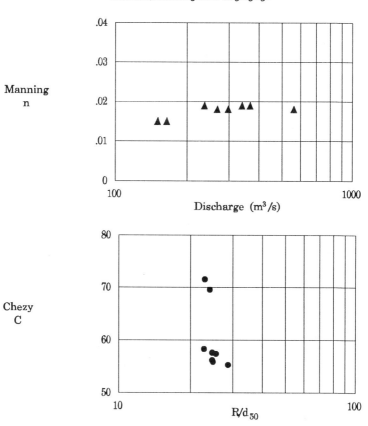

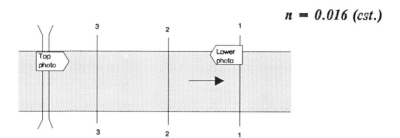

n = 0.016 (cst.)

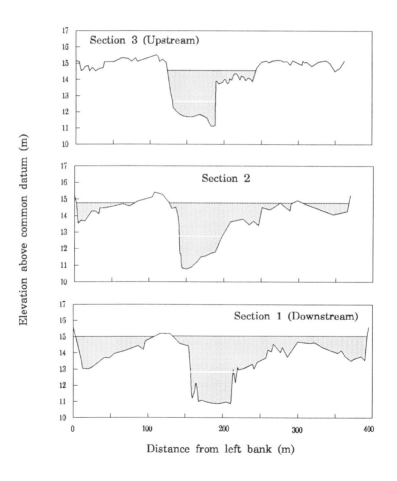

Plan (not to scale) and cross sections, Ngaruroro at Chesterhope Bridge.

n = 0.018

33379: Wanganui at Te Whaiau Canal.

Map reference:-	T19:355395 (Metric); N112:098974 (Yard).
Catchment area:-	221 km².
Period of record:-	September 1972 - May 1976.
Mean annual flood:-	19 m³/s.
Mean flow:-	10.9 m³/s.

Surveyed reach:-

Cross-sections:-	2 along a 281 m reach.
Manning's n range:-	0.018-0.022
Channel description:-	A man-made channel. Bed and banks consist of sand and gravel. Banks vegetated with grasses and flaxes.

Bed Surface Material

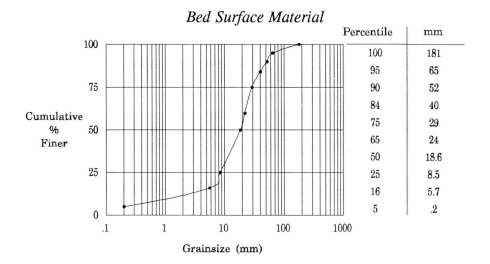

Percentile	mm
100	181
95	65
90	52
84	40
75	29
65	24
50	18.6
25	8.5
16	5.7
5	.2

View from top of reach looking downstream.

View upstream from middle of reach.

n = 0.018

Hydraulic Properties of Reach

Discharge	Water Surface Slope	Friction Slope	Area	Expansion	Hydraulic Radius	Mean Velocity	Manning n	Chezy C	Error
(m³/s)			(m²)	(%)	(m)	(m/s)			(%)
6.75*	0.00009	0.00010	16.2	20	0.81	0.42	0.021	46.9	14
6.79	0.00009	0.00009	17.0	22	0.84	0.40	0.022	44.8	14
10.7	0.00010	0.00011	19.8	21	0.94	0.55	0.018	54.3	14
12.4	0.00011	0.00013	21.2	20	0.99	0.59	0.019	52.8	13
13.0	0.00011	0.00012	22.5	19	1.03	0.58	0.020	51.2	13
13.5*	0.00011	0.00013	22.7	19	1.04	0.60	0.019	52.1	13

* Estimated from rating based on gaugings

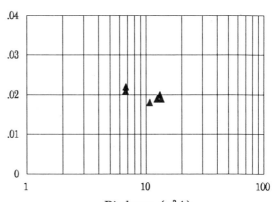

Manning n

Discharge (m³/s)

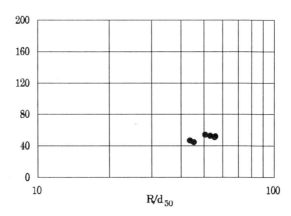

Chezy C

R/d$_{50}$

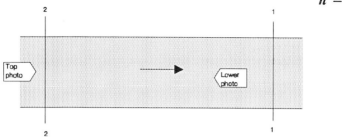

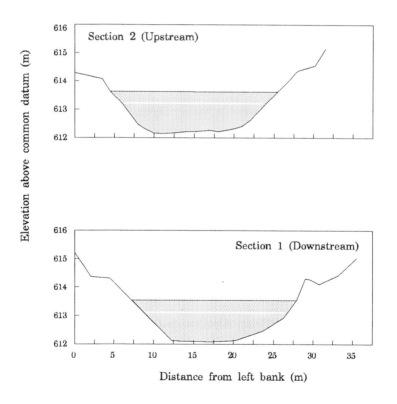

Plan (not to scale) and cross sections, Wanganui at Te Whaiau Canal.

n = 0.02

74347: Loganburn at Gorge (Downstream).

Map reference:-	H43:671233 (Metric); S144:627214 (Yard).
Catchment area:-	150 km².
Period of record:-	July 1980 - 1990.
Mean annual flood:-	37 m³/s.
Mean flow:-	1.64 m³/s.

Surveyed reach:-

Cross-sections:-	3 along a 45 m reach.
Manning's n range:-	0.020-0.039
Channel description:-	Bed is composed of gravel with small cobbles. Banks are lined with grass.

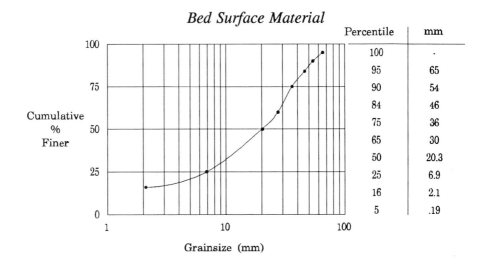

Bed Surface Material

Percentile	mm
100	.
95	65
90	54
84	46
75	36
65	30
50	20.3
25	6.9
16	2.1
5	.19

Grainsize (mm)

View upstream to top cross-section.

View downstream from middle of reach.

n = 0.02

Hydraulic Properties of Reach

Discharge	Water Surface Slope	Friction Slope	Area	Expansion	Hydraulic Radius	Mean Velocity	Manning n	Chezy C	Error
(m³/s)			(m²)	(%)	(m)	(m/s)			(%)
1.85*	0.00147	0.00157	3.45	39	0.36	0.56	0.039	21.6	10
2.05*	0.00153	0.00165	3.29	36	0.35	0.65	0.034	24.7	11
2.41*	0.00224	0.00249	3.46	48	0.36	0.73	0.038	22.5	9
2.92*	0.00196	0.00221	3.58	41	0.37	0.85	0.031	27.3	11
2.97*	0.00133	0.00110	4.14	-8	0.44	0.73	0.028	31.4	13
3.68*	0.00136	0.00162	3.66	31	0.37	1.03	0.022	39.4	15
4.06*	0.00164	0.00115	4.67	-20	0.47	0.88	0.024	36.7	15
4.20*	0.00156	0.00155	4.68	12	0.48	0.91	0.028	31.5	13
5.82*	0.00160	0.00211	4.24	29	0.43	1.40	0.020	43.9	19

* Estimated from rating based on gaugings

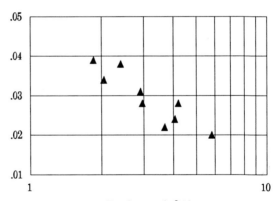

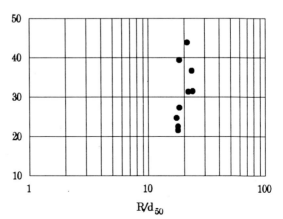

28

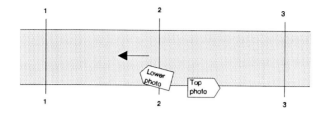

$n = 0.02$

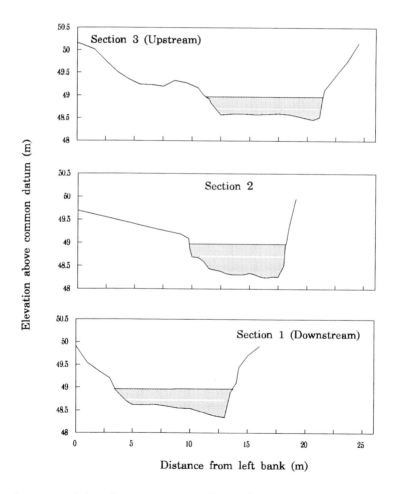

Plan (not to scale) and cross sections, Loganburn at Gorge (Downstream).

n = 0.024

33359: Wanganui at Wairehu Canal.

Map reference:-	T19:401398 (Metric); N112:148979 (Yard).
Catchment area:-	303 km².
Period of record:-	February 1971 - Present.
Mean annual flood:-	35 m³/s.
Mean flow:-	16.6 m³/s.

Surveyed reach:-

Cross-sections:-	2 along a 180 m reach.
Manning's n range:-	0.022-0.025
Channel description:-	A man-made channel. Bed and banks consist of uniformly-sized cobbles. Broom grows along the banks.

Bed Surface Material

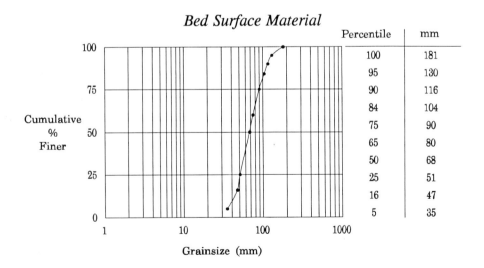

Percentile	mm
100	181
95	130
90	116
84	104
75	90
65	80
50	68
25	51
16	47
5	35

Grainsize (mm)

View from top of reach looking downstream.

View from bottom of reach looking upstream.

n = 0.024

Hydraulic Properties of Reach

Discharge	Water Surface Slope	Friction Slope	Area	Expansion	Hydraulic Radius	Mean Velocity	Manning n	Chezy C	Error
(m³/s)			(m²)	(%)	(m)	(m/s)			(%)
6.15	0.00029	0.00027	10.1	-10	0.83	0.61	0.024	40.7	13
6.22	0.00031	0.00029	10.3	-11	0.84	0.61	0.025	38.9	13
9.10*	0.00035	0.00032	12.6	-10	0.97	0.73	0.024	41.3	11
9.89	0.00034	0.00031	12.9	-9	0.98	0.77	0.023	44.2	11
13.9*	0.00043	0.00039	15.4	-9	1.11	0.91	0.023	43.7	11
14.9*	0.00048	0.00043	16.1	-10	1.14	0.93	0.024	42.4	11
15.9*	0.00052	0.00047	16.4	-10	1.15	0.97	0.024	42.0	11
26.5	0.00063	0.00053	20.9	-10	1.36	1.27	0.022	47.3	11
31.9	0.00069	0.00059	24.2	-10	1.50	1.32	0.024	44.5	11

* Estimated from rating based on gaugings

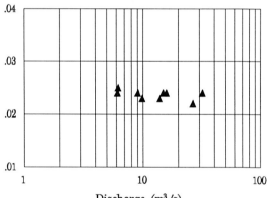

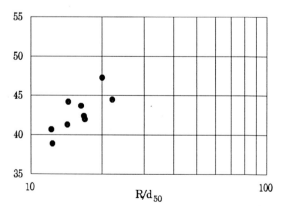

$$n = 0.024$$

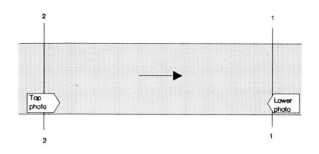

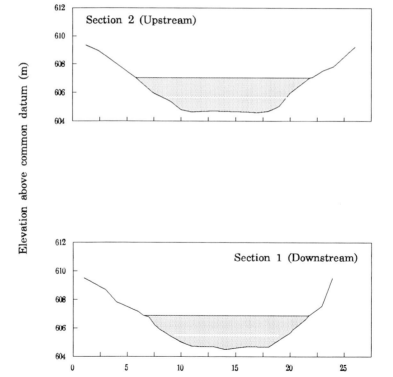

Plan (not to scale) and cross sections, Wanganui at Wairehu Canal.

n = 0.024

19716: Waipaoa at Kanakanaia Cableway.

Map reference:-	Y17:359923 (Metric); N089:273614 (Yard).
Catchment area:-	1582 km^2.
Period of record:-	November 1972 - Present.
Mean annual flood:-	1070 m^3/s.
Mean flow:-	34.2 m^3/s.

Surveyed reach:-

Cross-sections:-	3 along a 421 m reach.
Manning's n range:-	0.023-0.029
Channel description:-	Bed consists of fine gravel and silt. Both banks are lined with willow trees.

$n = 0.024$

View from top of reach looking downstream.

View from bottom of reach looking upstream.

n = 0.024

Hydraulic Properties of Reach

Discharge (m³/s)	Water Surface Slope	Friction Slope	Area (m²)	Expansion (%)	Hydraulic Radius (m)	Mean Velocity (m/s)	Manning n	Chezy C	Error[+] (%)
19.8	0.00070	0.00070	29.4	19	0.55	0.68	0.025	35.9	9
129*	0.00080	0.00087	91.1	30	1.10	1.44	0.023	44.7	10
415*	0.00076	0.00090	211	30	2.17	1.99	0.026	43.7	13
490*	0.00080	0.00092	244	26	2.47	2.02	0.028	41.7	13
672*	0.00071	0.00084	311	24	3.06	2.17	0.029	42.3	14

* Estimated from rating based on gaugings

+ Error includes additional 5% uncertainties in hydraulic radius and area values due to cross-section changes over the measurement period

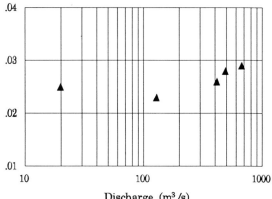

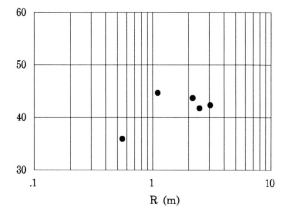

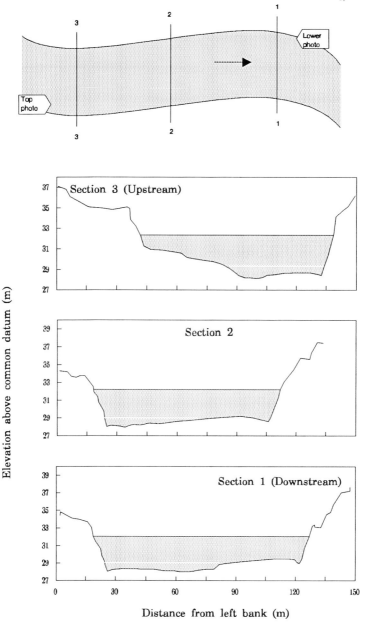

n = 0.024

Section 3 (Upstream)

Section 2

Section 1 (Downstream)

Elevation above common datum (m)

Distance from left bank (m)

Plan (not to scale) and cross sections, Waipaoa at Kanakanaia Cableway.

n = 0.025

79735: Waiau at Sunnyside.

Map reference:-	D44:934764 (Metric); S158:718733 (Yard).
Catchment area:-	6610 km².
Period of record:-	February 1972 - Present.
Mean annual flood:-	876 m³/s.
Mean flow:-	110 m³/s.

Surveyed reach:-

Cross-sections:-	3 along a 310 m reach.
Manning's n range:-	0.016-0.029
Channel description:-	Bed is composed of large cobbles, boulders, and bedrock. Banks are lined with native bush.

Bed Surface Material

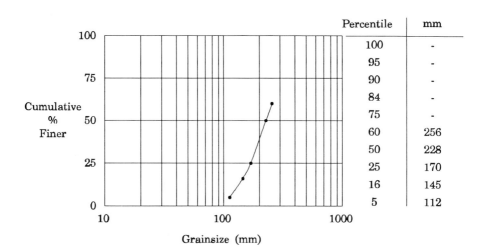

Percentile	mm
100	-
95	-
90	-
84	-
75	-
60	256
50	228
25	170
16	145
5	112

$n = 0.025$

View downstream from middle cross-section.

View upstream from bottom cross-section.

n = 0.025

Hydraulic Properties of Reach

Discharge	Water Surface Slope	Friction Slope	Area	Expansion	Hydraulic Radius	Mean Velocity	Manning n	Chezy C	Error
(m^3/s)			(m^2)	(%)	(m)	(m/s)			(%)
21.5*	0.00009	0.00011	66.7	152	1.03	0.37	0.028	35.1	13
21.6*	0.00009	0.00011	67.3	149	1.04	0.37	0.029	34.3	13
64.3*	0.00009	0.00013	104	82	1.47	0.66	0.023	46.8	14
103*	0.00016	0.00020	139	52	1.87	0.76	0.028	39.5	12
109*	0.00002	0.00008	132	59	1.79	0.86	0.016	70.7	28
188*	0.00007	0.00016	162	48	2.14	1.19	0.018	63.6	24
210*	0.00013	0.00020	179	39	2.33	1.19	0.021	54.2	20
405*	0.00018	0.00035	225	34	2.80	1.83	0.020	58.3	25
527*	0.00019	0.00038	256	30	3.08	2.08	0.020	60.6	29

* Estimated from rating based on gaugings

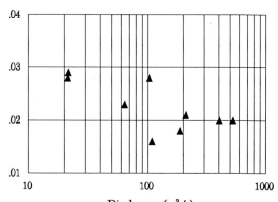

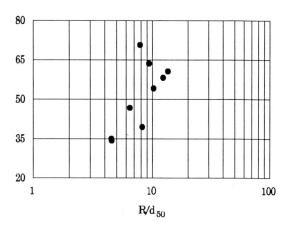

$n\ =\ 0.025$

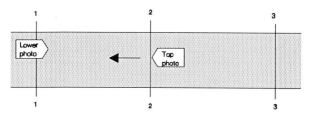

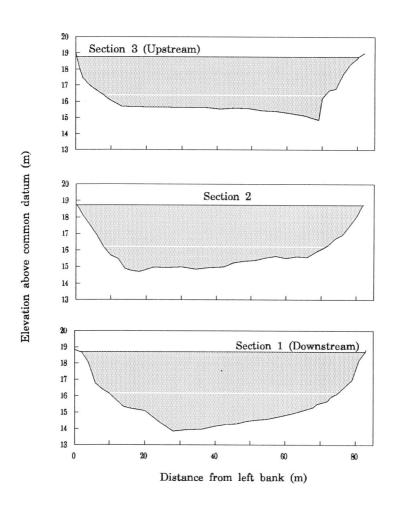

Plan (not to scale) and cross sections, Waiau at Sunnyside.

n = 0.026

71103: Hakataramea above Main Highway Bridge.

Map reference:-	I40:112062 (Metric); S118:125112 (Yard).
Catchment area:-	899 km².
Period of record:-	November 1963 - Present.
Mean annual flood:-	172 m³/s.
Mean flow:-	6.22 m³/s.

Surveyed reach:-

Cross-sections:-	5 along a 277 m reach.
Manning's n range:-	0.022-0.031
Channel description:-	Bed consists mainly of cobbles with some small boulders. Left bank is lined with willows; right bank is a tussock and Matagauri covered floodplain.

Bed Surface Material

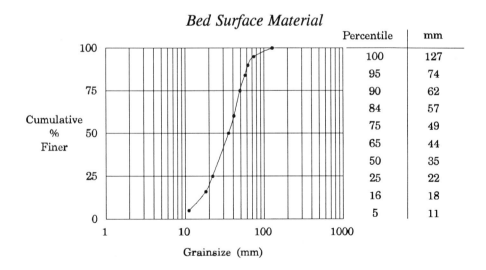

Percentile	mm
100	127
95	74
90	62
84	57
75	49
65	44
50	35
25	22
16	18
5	11

View downstream from top of reach.

View upstream from bottom of reach.

n = 0.026

Hydraulic Properties of Reach

Discharge	Water Surface Slope	Friction Slope	Area	Expansion	Hydraulic Radius	Mean Velocity	Manning n	Chezy C	Error
(m³/s)			(m²)	(%)	(m)	(m/s)			(%)
1.36	0.00088	0.00087	3.94	33	0.18	0.48	0.022	32.5	9
4.11	0.00084	0.00086	7.68	54	0.31	0.59	0.026	31.5	8
22.8	0.00148	0.00153	20.9	25	0.75	1.10	0.031	31.3	8

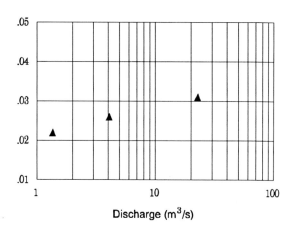

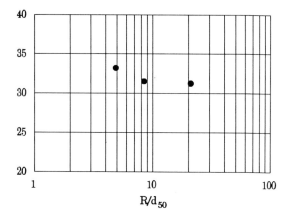

$$n = 0.026$$

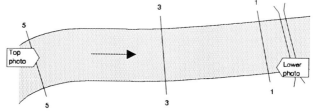

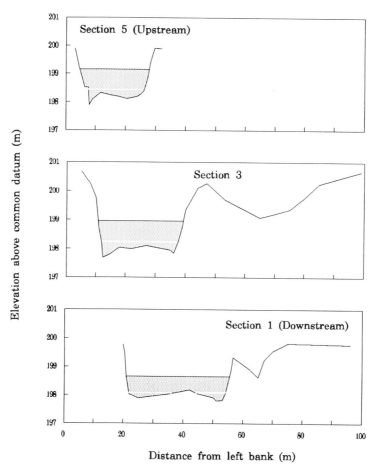

Plan (not to scale) and cross sections, Hakataramea above Main Highway Bridge.

n = 0.026

75214: Clutha at Lowburn.

Map reference:-	G41:121708 (Metric); S133:035744 (Yard).
Catchment area:-	6210 km^2.
Period of record:-	November 1967 - Present.
Mean annual flood:-	688 m^3/s.
Mean flow:-	296 m^3/s.

Surveyed reach:-

Cross-sections:- 3 along a 600 m reach.

Manning's n range:- 0.027-0.033

Channel description:- Bed material ranges from fine gravel to boulders approximately 300 mm in diameter. Banks are composed of cobble-boulder sized tailings (approximately 100 to 300 mm in diameter) with occasional sandy beaches. Banks have been cleared of vegetation.

View downstream from top cross-section.

View upstream from bottom cross-section.

n = 0.026

Hydraulic Properties of Reach

Discharge	Water Surface Slope	Friction Slope	Area	Expansion	Hydraulic Radius	Mean Velocity	Manning n	Chezy C	Error
(m³/s)			(m²)	(%)	(m)	(m/s)			(%)
126	0.00103	0.00101	102	-7	1.37	1.23	0.033	32.1	8
177*	0.00113	0.00110	121	-8	1.59	1.46	0.032	34.5	8
247*	0.00042	0.00040	148	1	1.89	1.67	0.027	41.6	11

* Estimated from rating based on gaugings

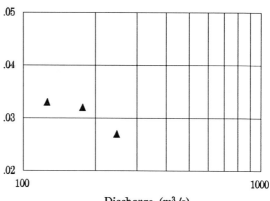

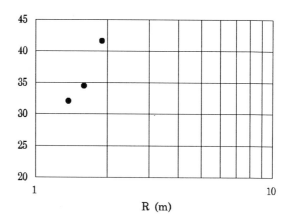

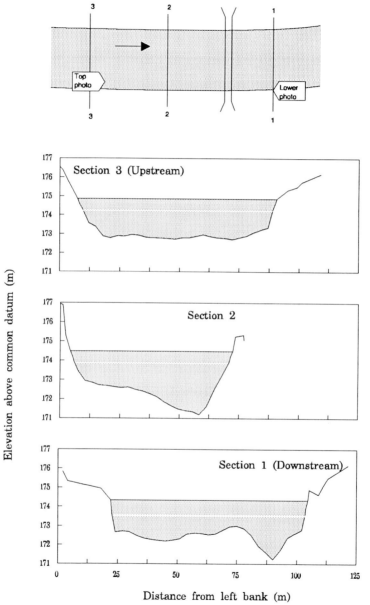

$n = 0.026$

Plan (not to scale) and cross sections, Clutha at Lowburn.

49

n = 0.027

93209: Maruia at Falls.

Map reference:- L29:478273 (Metric); S032:682596 (Yard).
Catchment area:- 980 km².
Period of record:- December 1963 - January 1990.
Mean annual flood:- 820 m³/s.
Mean flow:- 58.4 m³/s.

Surveyed reach:-
Cross-sections:- 5 along a 502 m reach.
Manning's n range:- 0.026-0.029
Channel description:- Bed consists predominantly of sand, with some silt and gravel. Banks are lined with overhanging trees, some dense scrub, and a little grass on the left bank only.

Bed Surface Material

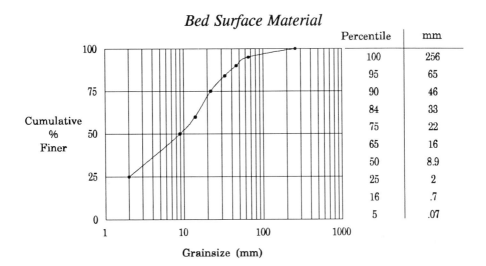

Percentile	mm
100	256
95	65
90	46
84	33
75	22
65	16
50	8.9
25	2
16	.7
5	.07

View upstream from top cross-section.

View downstream from top cross-section.

n = 0.027

Hydraulic Properties of Reach

Discharge	Water Surface Slope	Friction Slope	Area	Expansion	Hydraulic Radius	Mean Velocity	Manning n	Chezy C	Error
(m³/s)			(m²)	(%)	(m)	(m/s)			(%)
69.5*	0.00033	0.00033	87.2	22	1.58	0.81	0.027	39.0	9
141*	0.00044	0.00044	116	8	2.04	1.22	0.026	43.0	10
174*	0.00056	0.00055	127	3	2.22	1.36	0.028	40.9	9
290*	0.00080	0.00073	163	-6	2.69	1.78	0.029	41.0	10
511*	0.00136	0.00109	209	-17	3.12	2.45	0.028	42.6	12

* Estimated from rating based on gaugings

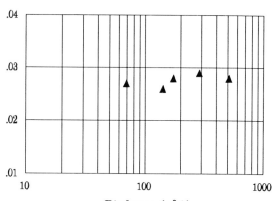

Manning n

Discharge (m³/s)

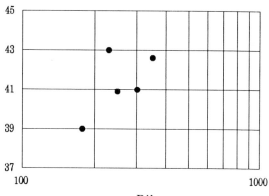

Chezy C

R/d_{50}

52

$n - 0.027$

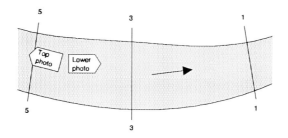

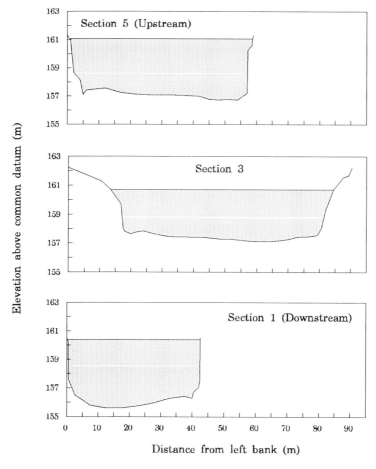

Plan (not to scale) and cross sections, Maruia at Falls.

53

n = 0.027 (est.)

8604: Orere at Bridge.

Map reference:-	S11:097682 (Metric); N043:745467 (Yard).
Catchment area:-	40.8 km².
Period of record:-	June 1978 - Present.
Mean annual flood:-	47 m³/s.
Mean flow:-	0.99 m³/s.

Surveyed reach:-

Cross-sections:-	3 along a 274 m reach.
Manning's n range:-	0.023-0.032
Channel description:-	Bed material ranges from smooth boulders down to small pebbles and some sand. Both banks are grassed to the water's edge, with some scrub on the right bank.

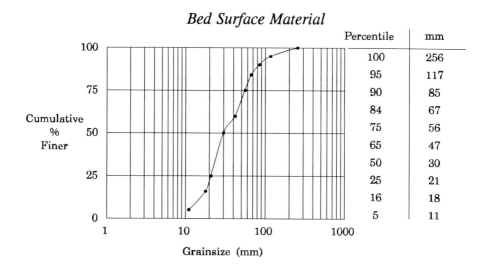

Bed Surface Material

Percentile	mm
100	256
95	117
90	85
84	67
75	56
65	47
50	30
25	21
16	18
5	11

View from cross-section 2 looking upstream.

View from cross-section 2 looking downstream.

n = 0.027 (est.)

Hydraulic Properties of Reach

Discharge	Water Surface Slope	Friction Slope	Area	Expansion	Hydraulic Radius	Mean Velocity	Manning n	Chezy C	Error
(m^3/s)			(m^2)	(%)	(m)	(m/s)			(%)
9.41*	0.00252	0.00253	8.14	5	0.64	1.16	0.032	28.9	8
11.6*	0.00212	0.00218	9.27	21	0.70	1.28	0.028	33.4	8
23.1*	0.00280	0.00272	11.5	-1	0.84	2.00	0.023	41.4	9
25.1*	0.00284	0.00275	12.5	-1	0.89	2.01	0.025	40.0	9
26.5*	0.00290	0.00280	13.4	-2	0.93	1.98	0.026	38.1	9
28.5*	0.00302	0.00288	14.1	-4	0.96	2.02	0.026	37.8	9
35.5*	0.00339	0.00303	17.9	-12	1.08	1.99	0.029	34.7	10
50.6*	0.00440	0.00319	19.5	-21	1.12	2.63	0.023	43.9	12

* Estimated from rating based on gaugings

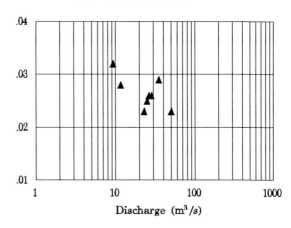

Manning n

Discharge (m³/s)

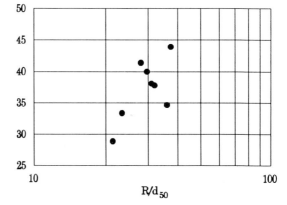

Chezy C

R/d₅₀
R/d$_{50}$

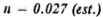

n − *0.027 (est.)*

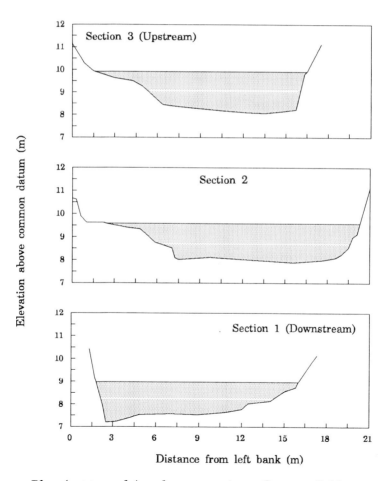

Plan (not to scale) and cross sections, Orere at Bridge.

n = *0.028*

91401: Grey at Dobson.

Map reference:-	J31:700601 (Metric); S044:818877 (Yard).
Catchment area:-	3830 km².
Period of record:-	July 1968 - Present.
Mean annual flood:-	3780 m³/s.
Mean flow:-	353 m³/s.

Surveyed reach:-

Cross-sections:-	5 along a 1518 m reach.
Manning's n range:-	0.025-0.031
Channel description:-	Bed is cobbled. Right bank is composed of bedrock, cobbles and soil, with sparse bush cover; left bank is composed of cobbles, silt and soil, and has thick bush cover with trees overhanging channel.

Bed Surface Material

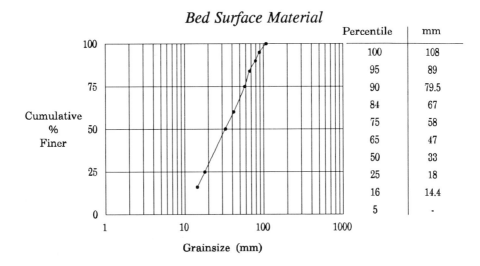

Percentile	mm
100	108
95	89
90	79.5
84	67
75	58
65	47
50	33
25	18
16	14.4
5	-

Grainsize (mm)

View upstream at top cross-section.

View downstream towards bottom of reach.

n = 0.028

Hydraulic Properties of Reach

Discharge	Water Surface Slope	Friction Slope	Area	Expansion	Hydraulic Radius	Mean Velocity	Manning n	Chezy C	Error
(m³/s)			(m²)	(%)	(m)	(m/s)			(%)
73.0	0.00088	0.00085	101	-22	0.67	0.78	0.031	29.8	8
116	0.00085	0.00082	122	-11	0.77	1.01	0.026	37.2	8
217	0.00069	0.00070	187	39	1.01	1.21	0.026	39.1	8
334	0.00087	0.00085	239	6	1.25	1.43	0.025	41.5	8
358	0.00088	0.00086	255	7	1.32	1.43	0.026	40.3	8
917	0.00105	0.00105	487	8	2.27	1.89	0.030	36.5	8
1110	0.00107	0.00107	501	7	2.33	2.22	0.026	44.2	8
3220*	0.00119	0.00122	959	18	3.96	3.37	0.026	48.4	9

* Estimated from rating based on gaugings

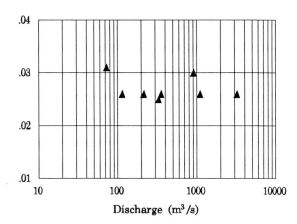

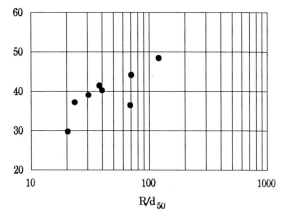

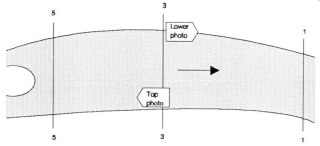

$n = 0.028$

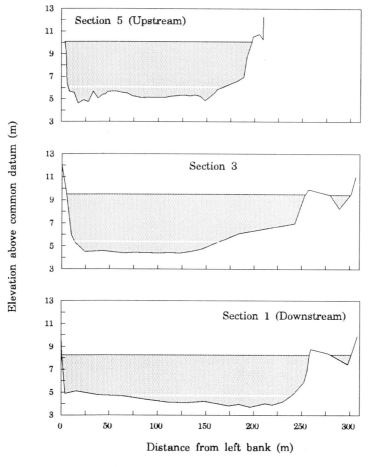

Plan (not to scale) and cross sections, Grey at Dobson.

n = 0.028

66602: Avon at Gloucester Street Bridge.

Map reference:-	M35:805419 (Metric); S084:003563 (Yard).
Catchment area:-	Artesian source.
Period of record:-	July 1980 - October 1989.
Mean annual flood:-	10 m³/s.
Mean flow:-	1.85 m³/s.

Surveyed reach:-

Cross-sections:-	3 along a 157 m reach.
Manning's n range:-	0.026-0.040
Channel description:-	Stream fed by artesian water and urban stormwater. Bed composed of angular cobbles and small boulders set in a muddy base. Left bank is grassed and steep; right bank has a lower grassy berm.

Bed Surface Material

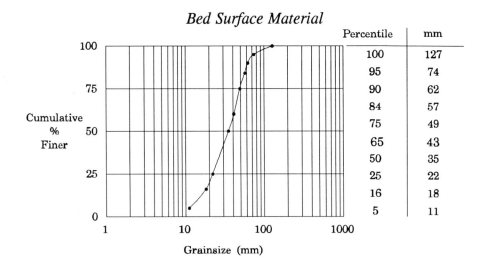

Percentile	mm
100	127
95	74
90	62
84	57
75	49
65	43
50	35
25	22
16	18
5	11

View upstream at cross-section 2.

$n = 0.028$

Hydraulic Properties of Reach

Discharge	Water Surface Slope	Friction Slope	Area	Expansion	Hydraulic Radius	Mean Velocity	Manning n	Chezy C	Error
(m³/s)			(m²)	(%)	(m)	(m/s)			(%)
1.83	0.00052	0.00048	4.38	-15	0.40	0.44	0.031	28.3	9
2.32	0.00051	0.00047	4.62	-5	0.42	0.53	0.026	33.3	10
3.74	0.00066	0.00058	6.11	-18	0.54	0.63	0.027	33.1	9
4.48*	0.00083	0.00073	6.84	-22	0.59	0.59	0.031	30.3	9
4.87*	0.00090	0.00078	7.16	-23	0.62	0.70	0.031	29.7	9
6.00*	0.00099	0.00086	8.06	-23	0.69	0.77	0.032	29.8	9
8.91*	0.00116	0.00098	10.0	-24	0.83	0.91	0.032	30.5	9
12.2*	0.00137	0.00115	12.2	-24	0.92	1.02	0.033	30.3	9
15.6*	0.00201	0.00171	14.9	-31	1.00	1.07	0.040	25.4	9
17.3*	0.00150	0.00125	15.6	-21	1.01	1.13	0.032	31.0	9

* Estimated from rating based on gaugings

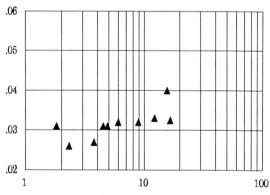

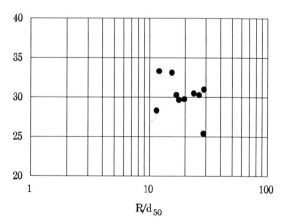

$n - 0.028$

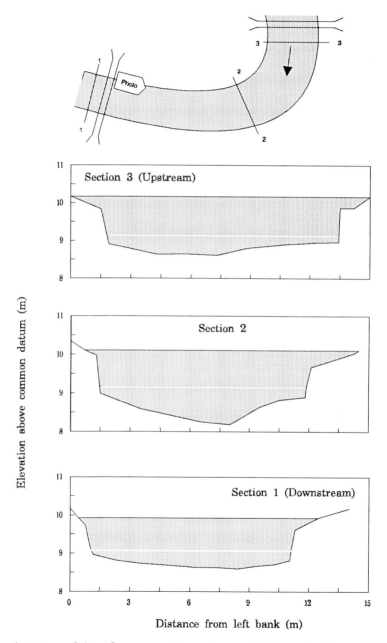

Plan (not to scale) and cross sections, Avon at Gloucester Street Bridge.

n = 0.028

79712: Monowai below Control Gates.

Map reference:-	C44:853751 (Metric); S158:629720 (Yard).
Catchment area:-	245 km².
Period of record:-	September 1976 - 1990.
Mean annual flood:-	39 m³/s.
Mean flow:-	16.8 m³/s.

Surveyed reach:-

Cross-sections:-	3 along a 120 m reach.
Manning's n range:-	0.026-0.030
Channel description:-	Bed is composed of coarse gravel and cobbles. Banks are lined with flax and overhanging willows.

$n = 0.028$

View downstream from top of reach.

View upstream from bottom of reach.

n = 0.028

Hydraulic Properties of Reach

Discharge	Water Surface Slope	Friction Slope	Area	Expansion	Hydraulic Radius	Mean Velocity	Manning n	Chezy C	Error
(m³/s)			(m²)	(%)	(m)	(m/s)			(%)
5.64	0.00097	0.00131	10.6	295	0.47	0.73	0.026	31.8	10
11.5*	0.00102	0.00164	15.0	245	0.60	1.00	0.027	32.4	11
14.1	0.00068	0.00098	17.1	110	0.67	0.91	0.026	35.3	11
19.2*	0.00072	0.00098	21.5	80	0.80	0.95	0.028	33.8	11
20.3*	0.00089	0.00114	22.0	73	0.81	0.98	0.030	31.8	10
20.3*	0.00077	0.00104	21.9	76	0.81	0.99	0.028	33.6	11
21.5*	0.00077	0.00106	22.4	77	0.83	1.02	0.029	33.6	11
21.7*	0.00069	0.00094	23.6	72	0.85	0.97	0.028	34.1	11
23.0*	0.00070	0.00094	24.6	67	0.87	0.99	0.028	34.2	11
23.1*	0.00068	0.00094	24.1	70	0.86	1.01	0.027	35.1	12
24.1*	0.00064	0.00091	24.7	70	0.88	1.03	0.027	35.9	12

* Estimated from rating based on gaugings

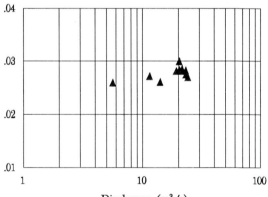

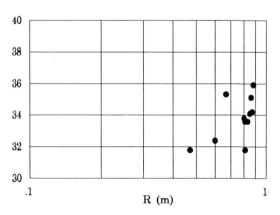

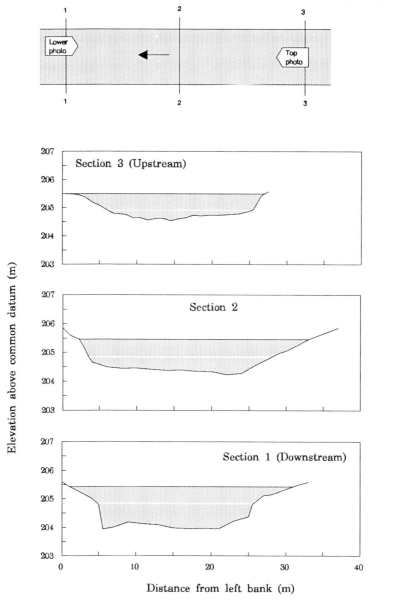

Plan (not to scale) and cross sections, Monowai below Control Gates.

n = 0.030

9140: Piako at Paeroa-Tahuna Bridge.

Map reference:- T13:318068 (Metric); N053:006802 (Yard).
Catchment area:- 534 km².
Period of record:- July 1972 - Present.
Mean annual flood:- 104 m³/s.
Mean flow:- 2.34 m³/s.

Surveyed reach:-
Cross-sections:- 3 along a 124 m reach.
Manning's n range:- 0.022-0.032
Channel description:- Bed consists of mud near banks and sand at centre of channel. Banks are lined with grass.

Bed Surface Material

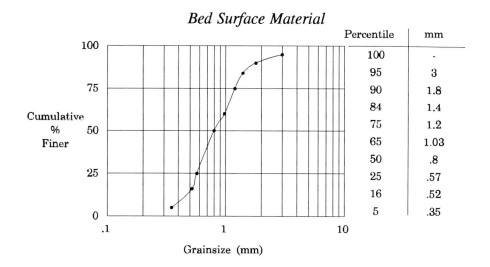

Percentile	mm
100	-
95	3
90	1.8
84	1.4
75	1.2
65	1.03
50	.8
25	.57
16	.52
5	.35

Grainsize (mm)

View upstream towards top cross-section.

View upstream towards middle cross-section.

n = 0.030

Hydraulic Properties of Reach

Discharge	Water Surface Slope	Friction Slope	Area	Expansion	Hydraulic Radius	Mean Velocity	Manning n	Chezy C	Error
(m³/s)			(m²)	(%)	(m)	(m/s)			(%)
2.82*	0.00011	0.00012	10.4	26	0.51	0.27	0.026	34.2	22
8.16*	0.00015	0.00016	21.1	13	0.95	0.39	0.032	31.5	18
11.9*	0.00006	0.00007	28.6	9	1.22	0.42	0.022	46.7	39
18.9*	0.00010	0.00010	42.2	6	1.66	0.45	0.031	35.3	27

* Estimated from rating based on gaugings

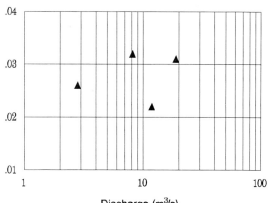

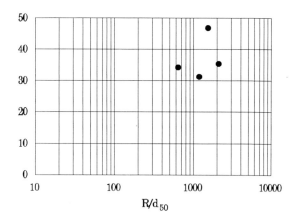

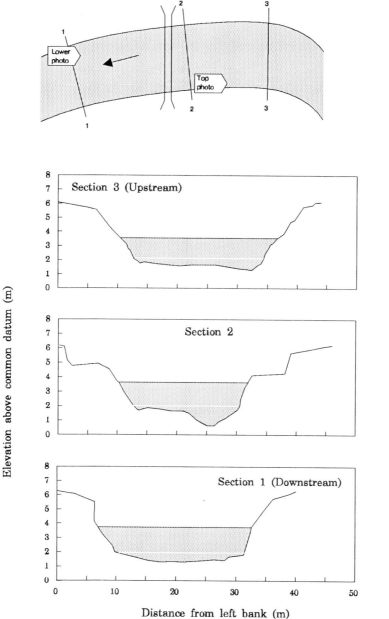

$n - 0.030$

Plan (not to scale) and cross sections, Piako at Paeroa-Tahuna Bridge.

73

n = 0.03

43402: Waikato at Ngaruawahia Cableway.

Map reference:-	S14:996924 (Metric); N056:658635 (Yard).
Catchment area:-	11400 km².
Period of record:-	May 1957 - February 1990.
Mean annual flood:-	777 m³/s.
Mean flow:-	328 m³/s.

Surveyed reach:-

Cross-sections:-	3 along a 3770 m reach.
Manning's n range:-	0.028-0.039
Channel description:-	Bed consists mainly of sand and gravel, with a little mud plus fragments of coal and pumice. Banks are lined with willows except for small patches of other trees and gorse.

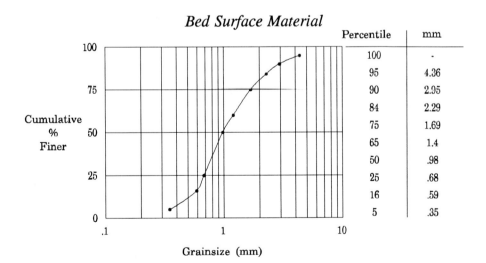

Bed Surface Material

Percentile	mm
100	-
95	4.36
90	2.95
84	2.29
75	1.69
65	1.4
50	.98
25	.68
16	.59
5	.35

Cumulative % Finer

Grainsize (mm)

View downstream at top cross-section.

View downstream at bottom cross-section.

n = 0.03

Hydraulic Properties of Reach

Discharge	Water Surface Slope	Friction Slope	Area	Expansion	Hydraulic Radius	Mean Velocity	Manning n	Chezy C	Error
(m³/s)			(m²)	(%)	(m)	(m/s)			(%)
237*	0.00013	0.00013	353	5	2.24	0.67	0.028	40.6	8
290*	0.00013	0.00013	410	8	2.50	0.71	0.029	40.2	8
448*	0.00015	0.00015	548	9	2.98	0.82	0.031	38.1	8
641*	0.00016	0.00016	726	13	3.74	0.89	0.035	35.9	8
738*	0.00017	0.00017	793	12	4.03	0.94	0.036	34.7	8
874*	0.00022	0.00022	855	9	4.32	1.03	0.039	32.9	9

* Estimated from rating based on gaugings

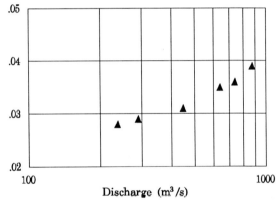

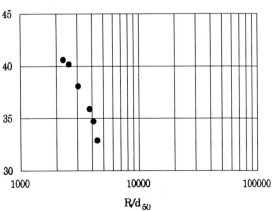

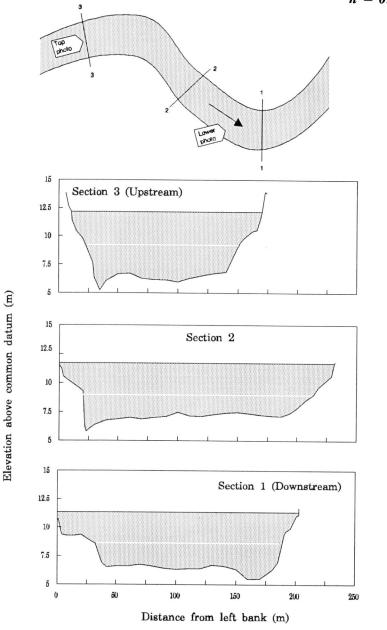

n = 0.03

Plan (not to scale) and cross sections, Waikato at Ngaruawahia Cableway.

n = 0.03

666000*: Heathcote at Sloan Terrace.

Map reference:-	M36:808376 (Metric); S084:008514 (Yard).
Catchment area:-	62.5 km².
Period of record:-	July 1989 - August 1989.
Mean annual flood:-	7 m³/s.
Mean flow:-	1.6 m³/s.

Surveyed reach:-

Cross-sections:-	2 along a 115 m reach.
Manning's n range:-	0.028-0.034
Channel description:-	Bed consists of angular cobbles and gravel intermixed with mud and silt. Banks are steep sided with long grass.

Bed Surface Material

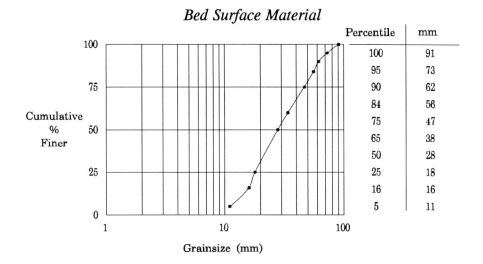

Percentile	mm
100	91
95	73
90	62
84	56
75	47
65	38
50	28
25	18
16	16
5	11

* River number only; no site number for this reach.

78

View from top cross-section looking downstream.

View of bottom cross-section looking upstream.

n = 0.03

Hydraulic Properties of Reach

Discharge	Water Surface Slope	Friction Slope	Area	Expansion	Hydraulic Radius	Mean Velocity	Manning n	Chezy C	Error
(m³/s)			(m²)	(%)	(m)	(m/s)			(%)
1.22*	0.00062	0.00064	2.76	29	0.37	0.45	0.029	29.1	9
1.74*	0.00059	0.00062	3.42	27	0.45	0.51	0.028	30.9	9
1.96*	0.00062	0.00065	3.72	24	0.48	0.53	0.029	30.2	9
2.12	0.00067	0.00070	3.94	19	0.50	0.54	0.031	28.9	9
2.94*	0.00064	0.00067	4.96	20	0.63	0.60	0.032	28.9	9
4.22*	0.00059	0.00062	6.58	16	0.76	0.65	0.032	29.6	10
4.83*	0.00058	0.00061	7.37	16	0.83	0.66	0.033	29.2	10
5.74	0.00040	0.00042	9.63	16	0.99	0.60	0.034	29.9	11
6.27	0.00031	0.00033	10.9	16	1.08	0.57	0.034	30.1	13
7.92	0.00038	0.00041	11.5	15	1.11	0.69	0.032	32.2	12
8.01	0.00031	0.00034	11.7	16	1.12	0.69	0.029	34.8	14
8.21*	0.00032	0.00035	11.9	16	1.27	0.69	0.032	32.8	13

* Estimated from rating based on gaugings

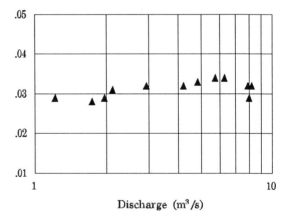

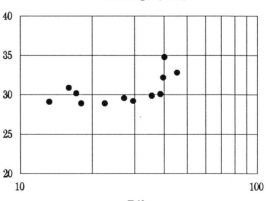

n — 0.03

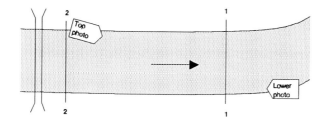

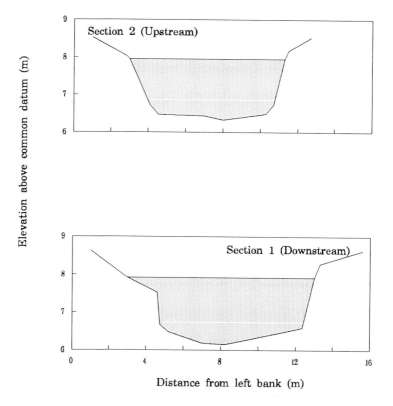

Plan (not to scale) and cross sections, Heathcote at Sloan Terrace.

n = 0.03

74319: Taieri below Patearoa Power Station.

Map reference:- H42:712380 (Metric); S145:675374 (Yard).
Catchment area:- 738 km².
Period of record:- April 1984 - 1990.
Mean annual flood:- 64 m³/s.
Mean flow:- 7.88 m³/s.

Surveyed reach:-
Cross-sections:- 3 along a 83 m reach.
Manning's n range:- 0.020-0.047
Channel description:- Bed comprises coarse gravel and small cobbles. Banks are lined with grass.

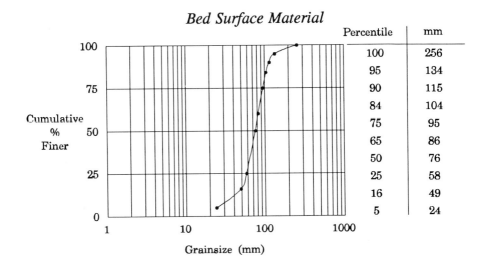

Bed Surface Material

Percentile	mm
100	256
95	134
90	115
84	104
75	95
65	86
50	76
25	58
16	49
5	24

View upstream from middle of reach.

View downstream from middle of reach.

n = 0.03

Hydraulic Properties of Reach

Discharge	Water Surface Slope	Friction Slope	Area	Expansion	Hydraulic Radius	Mean Velocity	Manning n	Chezy C	Error
(m³/s)			(m²)	(%)	(m)	(m/s)			(%)
0.78*	0.00034	0.00033	5.72	-21	0.30	0.15	0.046	17.2	14
1.24*	0.00048	0.00047	6.63	-18	0.34	0.20	0.047	17.4	11
6.13*	0.00099	0.00085	10.9	-30	0.54	0.58	0.031	29.1	10
6.66*	0.00099	0.00091	11.4	-9	0.57	0.59	0.033	27.7	9
9.10*	0.00090	0.00081	13.2	-10	0.65	0.69	0.029	31.9	10
11.3*	0.00094	0.00083	14.7	-11	0.71	0.77	0.028	33.1	11
11.8*	0.00087	0.00062	14.5	-26	0.72	0.83	0.023	41.7	13
12.0*	0.00086	0.00074	15.4	-12	0.74	0.78	0.027	34.9	11
12.3*	0.00095	0.00083	14.8	-10	0.74	0.84	0.026	35.9	11
18.7*	0.00095	0.00060	18.2	-25	0.84	1.05	0.020	48.7	16
20.4*	0.00093	0.00054	19.0	-25	0.88	1.09	0.029	52.4	18
21.2*	0.00114	0.00071	19.2	-27	0.87	1.13	0.021	47.3	15
27.1*	0.00135	0.00105	22.1	-15	0.96	1.23	0.025	40.0	14

* Estimated from rating based on gaugings

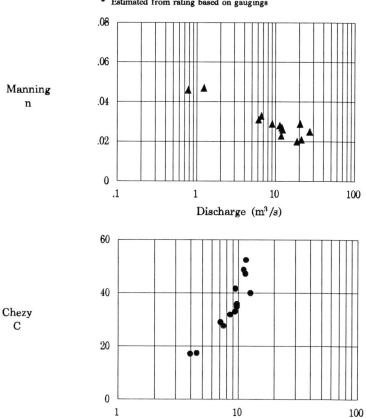

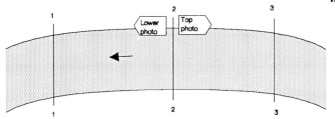

$n = 0.03$

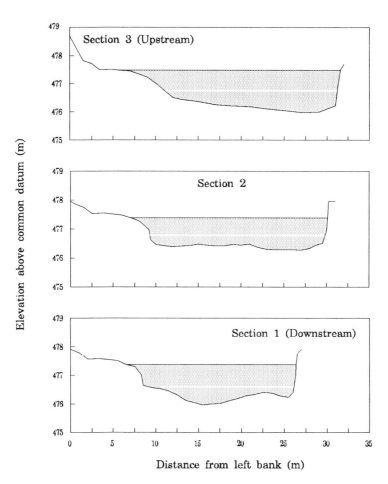

*Plan (not to scale) and cross sections, Taieri below Patearoa
Power Station.*

n = 0.03

74315: Taieri at MacAtamneys.

Map reference:-	H43:717292 (Metric); S145:679277 (Yard).
Catchment area:-	666 km².
Period of record:-	September 1977 - 1990.
Mean annual flood:-	121 m³/s.
Mean flow:-	3.13 m³/s.

Surveyed reach:-

Cross-sections:-	3 along a 90 m reach.
Manning's n range:-	0.026-0.033
Channel description:-	Bed comprises boulders and bedrock. Banks are grassy with tussocks. A few willows occur on each bank.

View upstream from middle of reach.

View downstream from middle of reach.

n = 0.03

Hydraulic Properties of Reach

Discharge	Water Surface Slope	Friction Slope	Area	Expansion	Hydraulic Radius	Mean Velocity	Manning n	Chezy C	Error
(m³/s)			(m²)	(%)	(m)	(m/s)			(%)
0.92	0.00111	0.00109	3.03	-9	0.18	0.31	0.033	22.8	9
0.96*	0.00136	0.00133	2.67	-13	0.18	0.36	0.030	24.7	8
7.40*	0.00136	0.00139	10.3	12	0.44	0.72	0.030	29.4	9
12.1*	0.00136	0.00137	14.0	6	0.56	0.87	0.029	31.4	9
12.7*	0.00123	0.00126	14.3	8	0.58	0.89	0.027	33.1	10
14.3*	0.00108	0.00111	15.3	10	0.61	0.94	0.026	36.1	10

* Estimated from rating based on gaugings

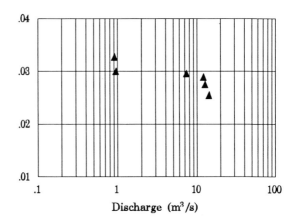

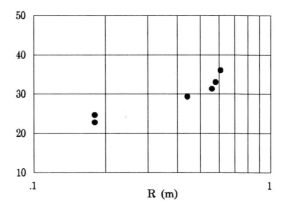

$$n = 0.03$$

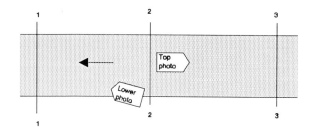

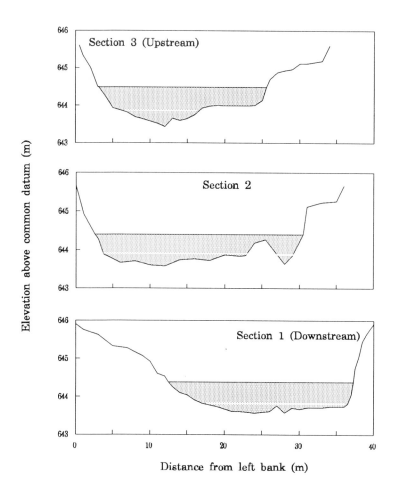

Plan (not to scale) and cross sections, Taieri at MacAtamneys.

n = 0.031

33316: Ongarue at Taringamotu.

Map reference:-	S18:043578 (Metric); N101:751165 (Yard).
Catchment area:-	1075 km².
Period of record:-	August 1962 - Present.
Mean annual flood:-	253 m³/s.
Mean flow:-	33.8 m³/s.

Surveyed reach:-

Cross-sections:-	3 along a 162 m reach.
Manning's n range:-	0.022-0.050
Channel description:-	Bed consists of gravel and cobbles. Farmland grasses line both banks.

Bed Surface Material

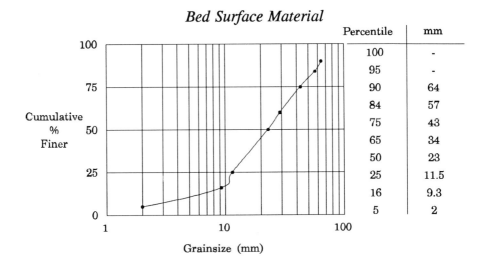

Percentile	mm
100	-
95	-
90	64
84	57
75	43
65	34
50	23
25	11.5
16	9.3
5	2

View from top cross-section looking downstream.

View from middle cross-section looking downstream.

_n = 0.031_

Hydraulic Properties of Reach

Discharge	Water Surface Slope	Friction Slope	Area	Expansion	Hydraulic Radius	Mean Velocity	Manning n	Chezy C	Error
(m³/s)			(m²)	(%)	(m)	(m/s)			(%)
10.5	0.00116	0.00099	25.8	-67	0.87	0.50	0.050	18.6	8
14.8	0.00050	0.00039	30.6	-50	1.01	0.52	0.034	28.3	10
18.7	0.00052	0.00039	33.2	-31	1.08	0.60	0.032	31.1	10
19.2	0.00049	0.00036	32.7	-46	1.06	0.63	0.028	34.4	10
35.1	0.00075	0.00054	41.7	-36	1.31	0.87	0.030	34.3	10
35.8	0.00065	0.00047	42.2	-34	1.32	0.87	0.028	36.8	11
41.7	0.00105	0.00081	44.5	-35	1.38	0.97	0.035	29.9	9
241	0.00027	0.00032	144	8	3.03	1.67	0.022	54.0	33

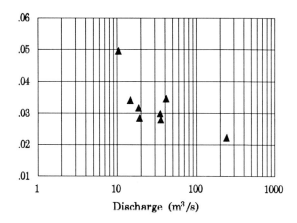

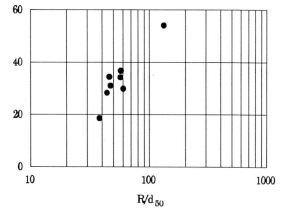

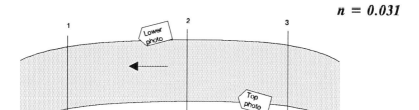

$$n = 0.031$$

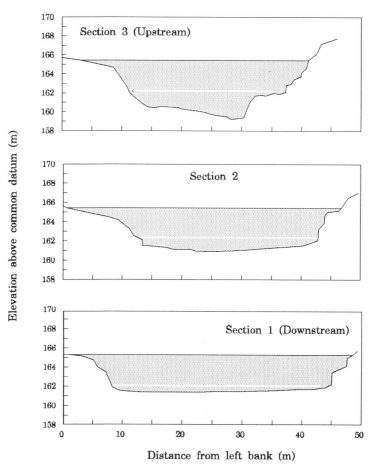

Plan (not to scale) and cross sections, Ongarue at Taringamotu.

n = 0.032

75232: Pomahaka at Burkes Ford.

Map reference:-	G45:314549 (Metric); S171:224472 (Yard).
Catchment area:-	1924 km².
Period of record:-	August 1961 - 1990.
Mean annual flood:-	408 m³/s.
Mean flow:-	26.9 m³/s.

Surveyed reach:-	
Cross-sections:-	3 along a 130 m reach.
Manning's n range:-	0.029-0.039
Channel description:-	Bed is composed of coarse gravel and large cobbles. Right bank is lined with willows; left bank has grassy slopes.

Bed Surface Material

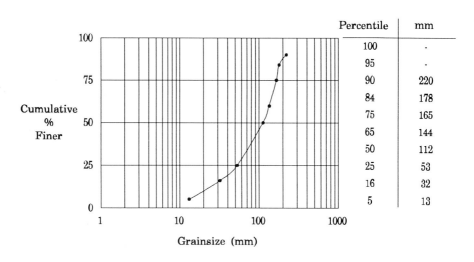

Percentile	mm
100	-
95	-
90	220
84	178
75	165
65	144
50	112
25	53
16	32
5	13

Grainsize (mm)

View downstream towards bottom cross-section.

View upstream from bottom cross-section.

_n = 0.032_

Hydraulic Properties of Reach

Discharge	Water Surface Slope	Friction Slope	Area	Expansion	Hydraulic Radius	Mean Velocity	Manning n	Chezy C	Error
(m³/s)			(m²)	(%)	(m)	(m/s)			(%)
4.58*	0.00011	0.00011	23.8	7	0.59	0.19	0.039	23.3	23
10.8	0.00014	0.00013	31.4	0	0.75	0.35	0.029	33.5	20
40.0	0.00045	0.00046	54.7	7	1.21	0.73	0.034	30.6	11
44.0	0.00047	0.00044	55.4	-2	1.23	0.79	0.031	33.4	12
62.9	0.00059	0.00051	65.8	-9	1.38	0.96	0.030	35.5	13
92.4	0.00057	0.00055	82.0	1	1.57	1.13	0.029	37.8	15
114	0.00065	0.00066	91.5	2	1.68	1.25	0.029	37.5	16

* Estimated from rating based on gaugings

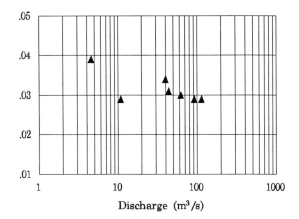

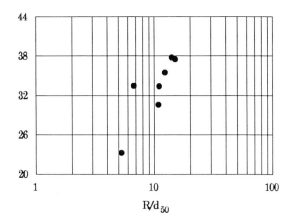

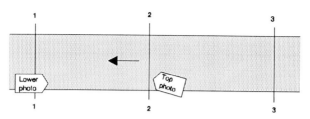

$n - 0.032$

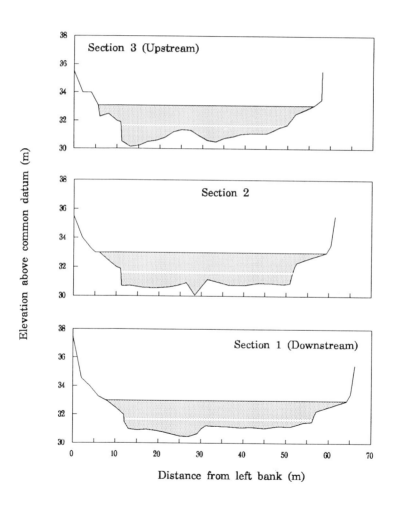

Plan (not to scale) and cross sections, Pomahaka at Burkes Ford.

n = 0.033 (est.)

19734: Waikohu at No. 1 Bridge.

Map reference:-	X17:083997 (Metric); N088:969686 (Yard).
Catchment area:-	26.4 km².
Period of record:-	July 1978 - Present.
Mean annual flood:-	32 m³/s.
Mean flow:-	0.91 m³/s.

Surveyed reach:-

Cross-sections:-	3 along a 112 m reach.
Manning's n range:-	0.032-0.033
Channel description:-	Bed material ranges from gravel to large cobbles, with some areas of exposed papa bedrock. Both banks are lined with overhanging trees.

Bed Surface Material

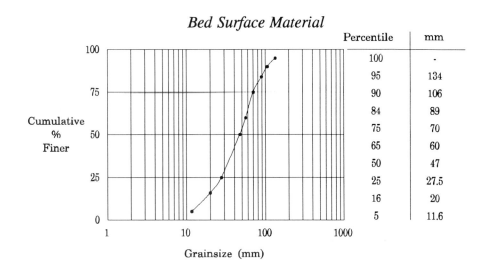

Percentile	mm
100	-
95	134
90	106
84	89
75	70
65	60
50	47
25	27.5
16	20
5	11.6

View downstream to cross-section 1.

View upstream towards cross-section 1.

n = *0.033 (est.)*

Hydraulic Properties of Reach

Discharge	Water Surface Slope	Friction Slope	Area	Expansion	Hydraulic Radius	Mean Velocity	Manning n	Chezy C	Error
(m³/s)			(m²)	(%)	(m)	(m/s)			(%)
18.7*	0.00268	0.00245	11.7	-6	1.02	1.59	0.032	31.1	10
19.1*	0.00268	0.00248	12.1	-5	1.04	1.58	0.033	30.1	10

* Estimated from rating based on gaugings

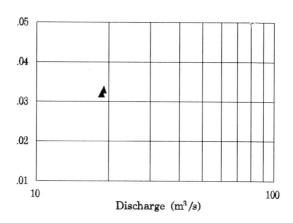

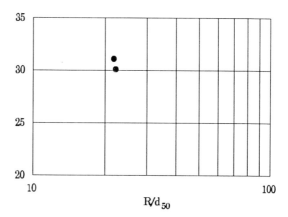

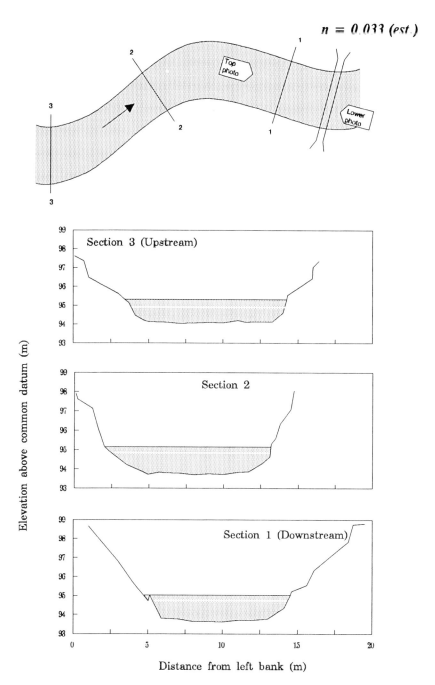

$n = 0.033 \ (est.)$

Plan (not to scale) and cross sections, Waikohu at No. 1 Bridge.

101

n = 0.033 *(est.)*

18913: Mangaheia at Willowbank.

Map reference:-	Y17:634066 (Metric); N089:569778 (Yard).
Catchment area:-	40.3 km².
Period of record:-	November 1988 - Present.
Mean annual flood:-	80 m³/s.
Mean flow:-	1.21 m³/s.

Surveyed reach:-

Cross-sections:-	3 along a 172 m reach.
Manning's n range:-	0.029-0.044
Channel description:-	Bed material ranges from gravel to large boulders. Banks are grassed with overhanging willows on both sides.

Bed Surface Material

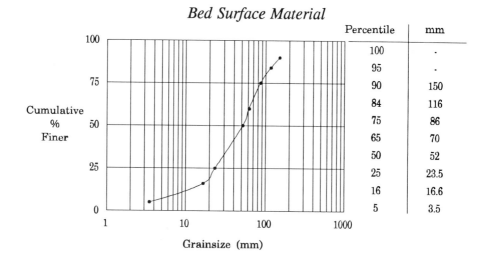

Percentile	mm
100	.
95	.
90	150
84	116
75	86
65	70
50	52
25	23.5
16	16.6
5	3.5

$n = 0.033$ *(est.)*

View downstream from middle of reach.

View from bottom of reach looking upstream.

n = 0.033 (est.)

Hydraulic Properties of Reach

Discharge	Water Surface Slope	Friction Slope	Area	Expansion	Hydraulic Radius	Mean Velocity	Manning n	Chezy C	Error
(m^3/s)			(m^2)	(%)	(m)	(m/s)			(%)
7.0*	0.00402	0.00378	9.17	-50	0.63	0.86	0.044	20.2	10
20.0*	0.00236	0.00215	14.4	-11	0.90	1.41	0.029	33.5	9
25.0*	0.00249	0.00224	16.5	-9	1.00	1.53	0.029	33.9	9
60.0*	0.00322	0.00284	29.5	-5	1.44	2.06	0.032	33.5	8

* Estimated from rating based on gaugings

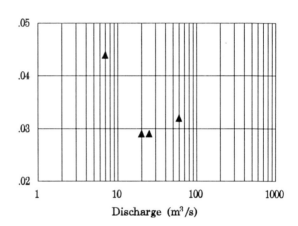

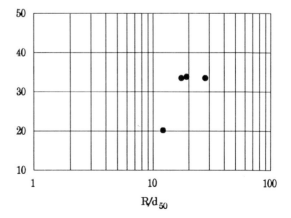

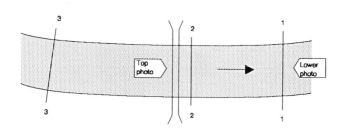

$n = 0.033 \ (est.)$

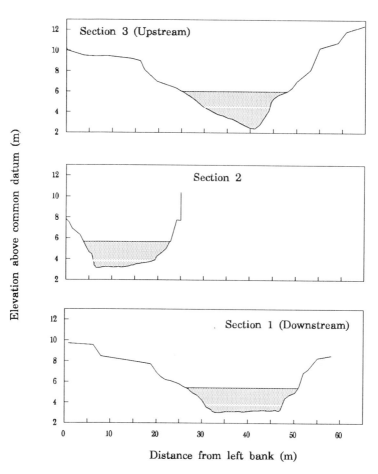

Plan (not to scale) and cross sections, Mangaheia at Willowbank.

n = 0.033

68521: Oakden Canal at Oakden Culvert.

Map reference:-	K34:832749 (Metric); S065:946943 (Yard).
Catchment area:-	NA.
Period of record:-	December 1979 - Present.
Mean annual flood:-	36 m³/s.
Mean flow:-	7.32 m³/s.

Surveyed reach:-

Cross-sections:-	2 along a 635 m reach.
Manning's n range:-	0.027-0.037
Channel description:-	A large man-made canal. Bed and banks are of compacted gravel and boulders, with areas of silt and weed growth.

Bed Surface Material

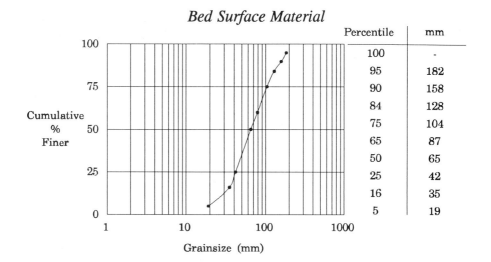

Percentile	mm
100	-
95	182
90	158
84	128
75	104
65	87
50	65
25	42
16	35
5	19

$n = 0.033$

View downstream at top cross-section.

View upstream from bottom cross-section.

n = 0.033

Hydraulic Properties of Reach

Discharge	Water Surface Slope	Friction Slope	Area	Expansion	Hydraulic Radius	Mean Velocity	Manning n	Chezy C	Error
(m³/s)			(m²)	(%)	(m)	(m/s)			(%)
4.90	0.00001	0.00001	35.5	-32	1.77	0.14	0.037	29.6	28
8.00	0.00003	0.00002	36.7	-32	1.81	0.23	0.031	35.3	18
13.9	0.00006	0.00005	38.5	-32	1.85	0.37	0.030	37.2	11
20.5	0.00010	0.00008	39.9	-32	1.89	0.53	0.027	41.8	10

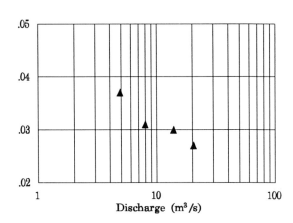

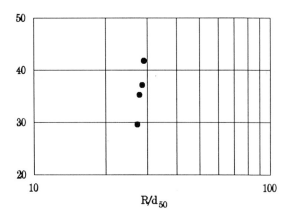

$n = 0.033$

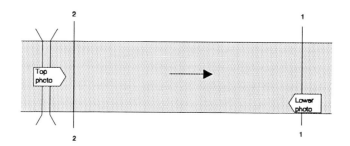

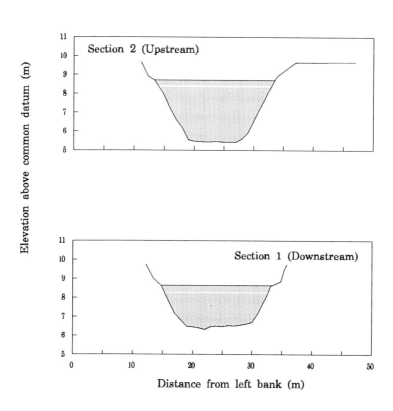

Plan (not to scale) and cross sections, Oakden Canal at Oakden Culvert.

n = 0.034

64616: Waiau Water Race at Lateral 2.

Map reference:-	N32:993322 (Metric); S054:226546 (Yard).
Catchment area:-	NA.
Period of record:-	September 1981 - May 1984.
Mean annual flood:-	NA.
Mean flow:-	0.72 m³/s.

Surveyed reach:-

Cross-sections:-	3 along a 120 m reach.
Manning's n range:-	0.027-0.045
Channel description:-	Man-made, lined irrigation channel. Bed is mixed cobbles and gravel. Banks are covered with long grass.

Bed Surface Material

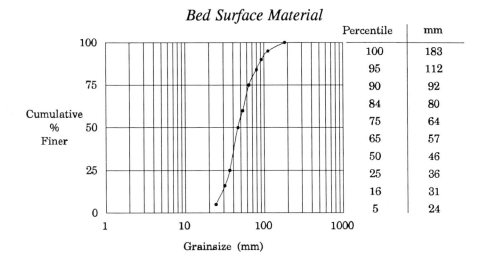

Percentile	mm
100	183
95	112
90	92
84	80
75	64
65	57
50	46
25	36
16	31
5	24

View downstream at top cross-section.

View upstream at middle cross-section.

n = 0.034

Hydraulic Properties of Reach

Discharge	Water Surface Slope	Friction Slope	Area	Expansion	Hydraulic Radius	Mean Velocity	Manning n	Chezy C	Error
(m³/s)			(m²)	(%)	(m)	(m/s)			(%)
0.33*	0.00459	0.00455	0.71	-13	0.18	0.47	0.045	16.7	8
0.68	0.00473	0.00462	0.89	-20	0.22	0.77	0.032	24.5	8
1.22*	0.00475	0.00460	1.19	-15	0.28	1.03	0.028	28.8	8
1.40*	0.00466	0.00459	1.34	-7	0.30	1.05	0.029	28.7	8
1.90*	0.00476	0.00461	1.57	-12	0.35	1.21	0.027	30.6	8
2.12*	0.00483	0.00468	1.66	-11	0.36	1.28	0.027	31.2	8
2.22	0.00478	0.00470	1.75	-6	0.37	1.27	0.028	30.8	8
2.52	0.00484	0.00468	1.91	-10	0.38	1.33	0.027	31.5	8

* Estimated from rating based on gaugings

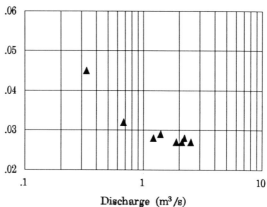

Discharge (m³/s)

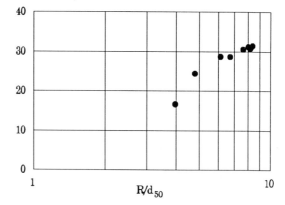

R/d_{50}

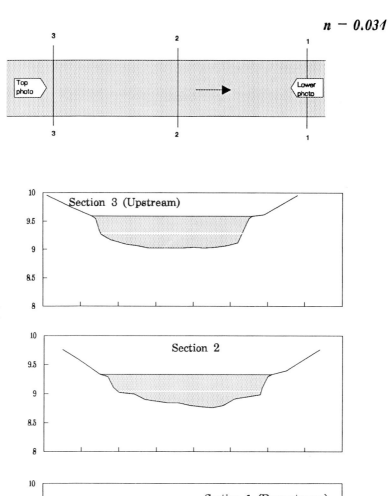

$n - 0.031$

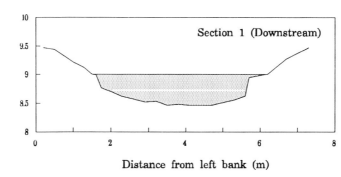

Plan (not to scale) and cross sections, Waiau Water Race at Lateral 2.

n = 0.035 (est.)

15901: Waioeka at Gorge Cableway.

Map reference:-   W16:877220 (Metric); N078:737923 (Yard).
Catchment area:-   652 km².
Period of record:-   March 1958 - Present.
Mean annual flood:-   661 m³/s.
Mean flow:-   32.2 m³/s.

Surveyed reach:-
Cross-sections:-   3 along a 312 m reach.
Manning's n range:-   0.033-0.036
Channel description:-   Bed consists mainly of cobbles and gravel. Both banks are heavily bushed, with large trees overhanging the left bank.

Bed Surface Material

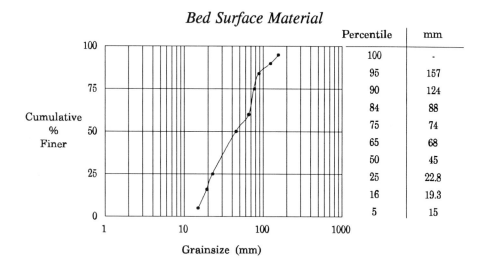

Percentile	mm
100	-
95	157
90	124
84	88
75	74
65	68
50	45
25	22.8
16	19.3
5	15

View from top of reach looking downstream.

View from bottom of reach looking upstream.

115

n = 0.035 (est.)

Hydraulic Properties of Reach

Discharge	Water Surface Slope	Friction Slope	Area	Expansion	Hydraulic Radius	Mean Velocity	Manning n	Chezy C	Error
(m³/s)			(m²)	(%)	(m)	(m/s)			(%)
249*	0.00209	0.00268	86.8	216	1.94	3.66	0.033	33.3	14
638*	0.00144	0.00305	191	99	3.52	3.61	0.035	35.2	15
662*	0.00165	0.00295	197	92	3.61	3.59	0.036	34.1	15
717*	0.00137	0.00515	214	88	3.84	3.57	0.035	35.4	16

* Estimated from rating based on gaugings

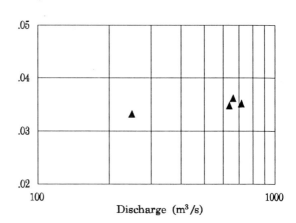

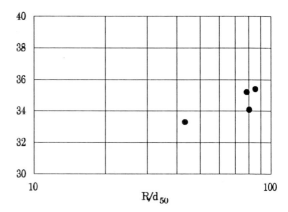

116

$$n - 0.035 \text{ (est.)}$$

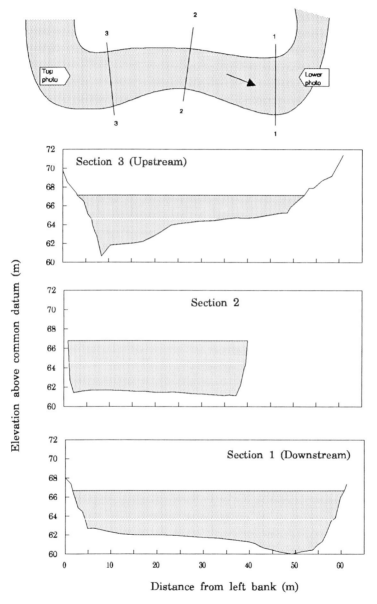

Plan (not to scale) and cross sections, Waioeka at Gorge
Cableway.

n = 0.036

33107: Whangaehu at Karioi.

Map reference:-	S21:218864 (Metric); N131:965389 (Yard).
Catchment area:-	492 km².
Period of record:-	November 1962 - 1989.
Mean annual flood:-	92.1 m³/s.
Mean flow:-	15.1 m³/s.

Surveyed reach:-

Cross-sections:-	5 along a 160 m reach.
Manning's n range:-	0.028-0.041
Channel description:-	Bed material is mainly sand and cobbles. Left bank has some flax overhanging the channel.

Bed Surface Material

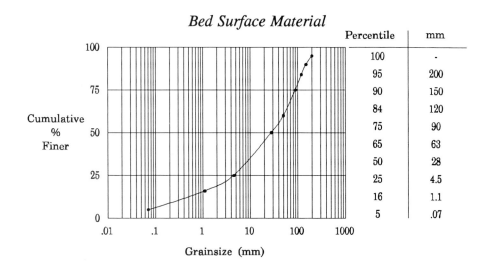

Percentile	mm
100	-
95	200
90	150
84	120
75	90
65	63
50	28
25	4.5
16	1.1
5	.07

Grainsize (mm)

$n = 0.036$

View upstream from cross-section 3.

View downstream from cross-section 3.

n = 0.036

Hydraulic Properties of Reach

Discharge	Water Surface Slope	Friction Slope	Area	Expansion	Hydraulic Radius	Mean Velocity	Manning n	Chezy C	Error
(m³/s)			(m²)	(%)	(m)	(m/s)			(%)
8.20*	0.00193	0.00198	10.3	40	0.68	0.81	0.041	22.7	8
10.6*	0.00196	0.00202	11.5	36	0.74	0.94	0.038	25.0	8
11.5*	0.00201	0.00207	11.9	33	0.76	0.97	0.038	25.1	8
14.1*	0.00192	0.00198	13.1	32	0.82	1.09	0.035	27.7	8
25.2*	0.00212	0.00214	17.2	19	1.02	1.47	0.031	32.3	9
61.0*	0.00306	0.00268	27.4	1	1.39	2.25	0.028	37.7	11

* Estimated from rating based on gaugings

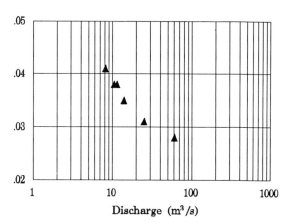

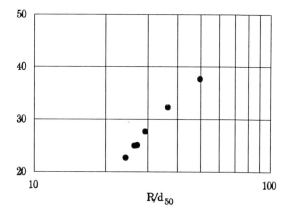

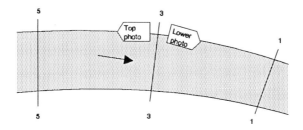

$n = 0.036$

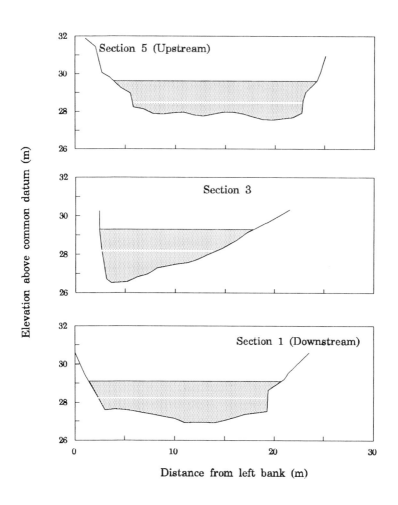

Plan (not to scale) and cross sections, Whangaehu at Karioi.

n = 0.037

75290: Cardrona at Albert-town.

Map reference:-	F40:079064 (Metric); S115:995134 (Yard).
Catchment area:-	346 km².
Period of record:-	September 1978 - Present.
Mean annual flood:-	55.4 m³/s.
Mean flow:-	3.08 m³/s.

Surveyed reach:-

Cross-sections:-	3 along a 150 m reach.
Manning's n range:-	0.033-0.037
Channel description:-	Bed material ranges from cobbles to boulders. Banks are grassed, with large willows overhanging both banks.

Bed Surface Material

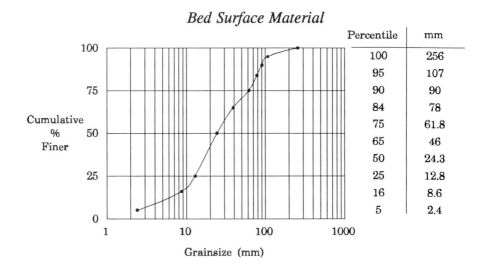

Percentile	mm
100	256
95	107
90	90
84	78
75	61.8
65	46
50	24.3
25	12.8
16	8.6
5	2.4

$$n = 0.037$$

View from middle cross-section looking downstream.

View from bottom cross-section looking upstream.

n = 0.037

Hydraulic Properties of Reach

Discharge	Water Surface Slope	Friction Slope	Area	Expansion	Hydraulic Radius	Mean Velocity	Manning n	Chezy C	Error
(m³/s)			(m²)	(%)	(m)	(m/s)			(%)
2.57*	0.00806	0.00806	2.60	4	0.28	0.99	0.037	21.5	8
3.09*	0.00829	0.00820	2.98	-6	0.30	1.04	0.037	21.9	8
3.60*	0.00837	0.00821	3.28	-14	0.32	1.11	0.037	22.5	8
4.24*	0.00835	0.00812	3.59	-13	0.34	1.19	0.035	23.5	8
5.10*	0.00835	0.00816	4.02	-13	0.37	1.28	0.035	24.4	8
5.83*	0.00838	0.00812	4.35	-16	0.39	1.36	0.034	25.4	8
6.60*	0.00845	0.00810	4.71	-20	0.40	1.43	0.033	26.2	8
7.61*	0.00851	0.00812	5.35	-23	0.42	1.46	0.033	25.9	8

* Estimated from rating based on gaugings

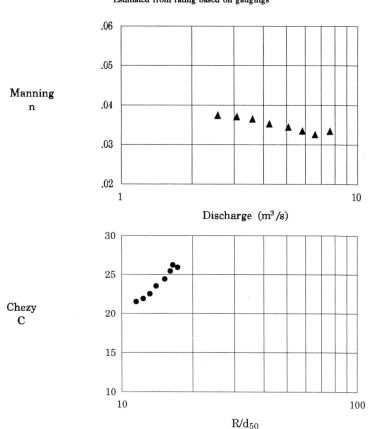

124

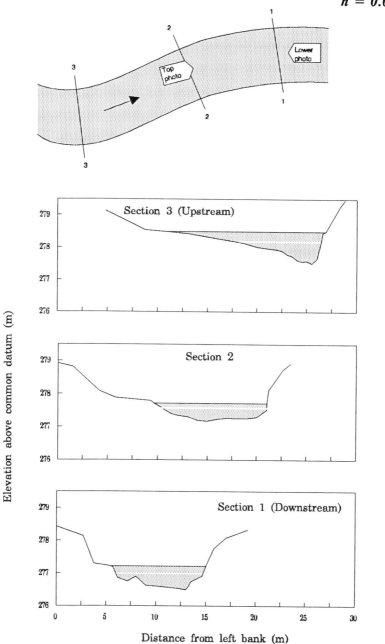

Plan (not to scale) and cross sections, Cardrona at Albert-town.

n = 0.037

29808: Hutt at Kaitoke.

Map reference:-	S26:942150 (Metric); N161:716507 (Yard).
Catchment area:-	88.8 km².
Period of record:-	December 1967 - December 1988.
Mean annual flood:-	280 m³/s.
Mean flow:-	7.6 m³/s.

Surveyed reach:-

Cross-sections:-	3 along a 55 m reach.
Manning's n range:-	0.021-0.053
Channel description:-	Bed consists of boulders, cobbles, and gravel. Banks are lined with overhanging trees and scrub.

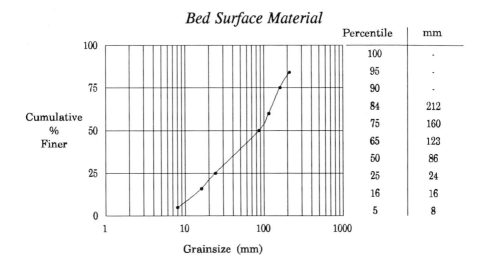

Bed Surface Material

Percentile	mm
100	-
95	-
90	-
84	212
75	160
65	123
50	86
25	24
16	16
5	8

View downstream at middle of reach.

View upstream from bottom cross-section.

n = 0.037

Hydraulic Properties of Reach

Discharge	Water Surface Slope	Friction Slope	Area	Expansion	Hydraulic Radius	Mean Velocity	Manning n	Chezy C	Error
(m³/s)			(m²)	(%)	(m)	(m/s)			(%)
3.53*	0.00133	0.00097	7.03	-51	0.26	0.55	0.021	36.6	11
8.38*	0.00456	0.00359	11.8	-57	0.42	0.79	0.039	21.8	9
8.69*	0.00318	0.00291	13.8	-31	0.48	0.65	0.048	18.1	8
17.2*	0.00376	0.00326	20.3	-31	0.67	0.88	0.047	19.6	9
77.2*	0.00805	0.00637	44.2	-26	1.32	1.77	0.053	19.7	8
104*	0.00752	0.00590	50.5	-19	1.45	2.08	0.046	22.9	8

* Estimated from rating based on gaugings

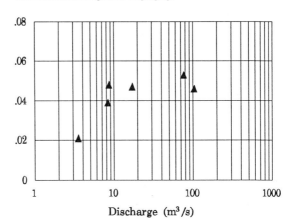

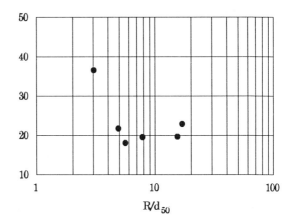

$n = 0.037$

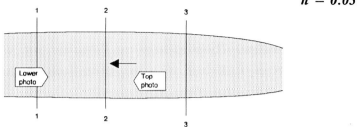

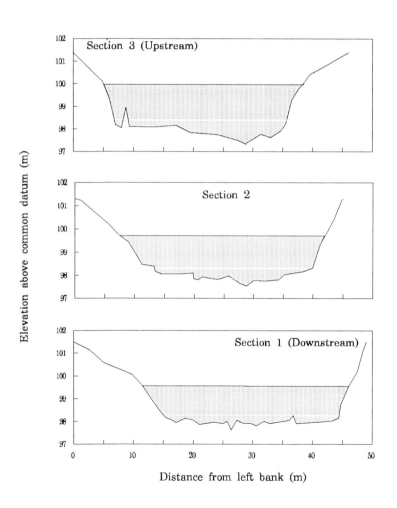

Plan (not to scale) and cross sections, Hutt at Kaitoke.

n = 0.037

62105: Clarence at Jollies.

Map reference:-	N31:023611 (Metric); S047:265862 (Yard).
Catchment area:-	440 km².
Period of record:-	April 1958 - Present.
Mean annual flood:-	194 m³/s.
Mean flow:-	15.1 m³/s.

Surveyed reach:-

Cross-sections:-	3 along a 193 m reach.
Manning's n range:-	0.025-0.042
Channel description:-	Bed consists of small boulders and gravel with some patches of exposed bedrock. Banks are lined with tussock and grasses.

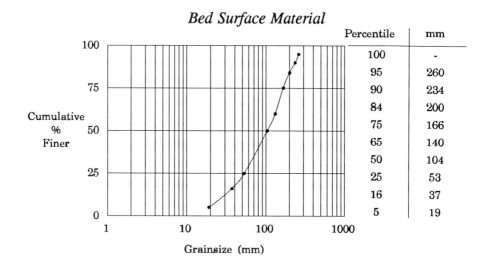

Bed Surface Material

Percentile	mm
100	-
95	260
90	234
84	200
75	166
65	140
50	104
25	53
16	37
5	19

$n = 0.037$

View upstream towards top cross-section.

View downstream towards middle and lower cross-sections.

n = 0.037

Hydraulic Properties of Reach

Discharge	Water Surface Slope	Friction Slope	Area	Expansion	Hydraulic Radius	Mean Velocity	Manning n	Chezy C	Error
(m³/s)			(m²)	(%)	(m)	(m/s)			(%)
7.62	0.00332	0.00327	11.8	-19	0.38	0.66	0.042	19.9	11
12.4*	0.00310	0.00302	14.7	-19	0.46	0.85	0.036	24.2	10
17.5*	0.00312	0.00302	17.4	-17	0.53	1.02	0.034	26.6	8
18.1	0.00322	0.00312	19.0	-19	0.58	0.96	0.038	23.6	8
24.0*	0.00318	0.00305	20.7	-17	0.62	1.17	0.033	28.0	8
39.7*	0.00325	0.00305	26.6	-16	0.77	1.50	0.030	32.1	8
64.8*	0.00320	0.00296	35.0	-13	0.99	1.86	0.028	35.2	8
106*	0.00322	0.00287	44.6	-11	1.23	2.38	0.025	40.9	8
120*	0.00332	0.00295	50.8	-12	1.38	2.38	0.028	38.0	8

* Estimated from rating based on gaugings

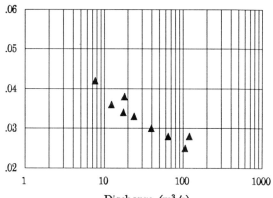

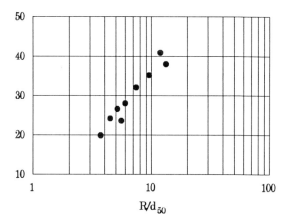

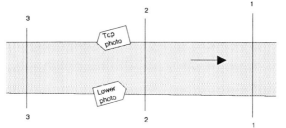

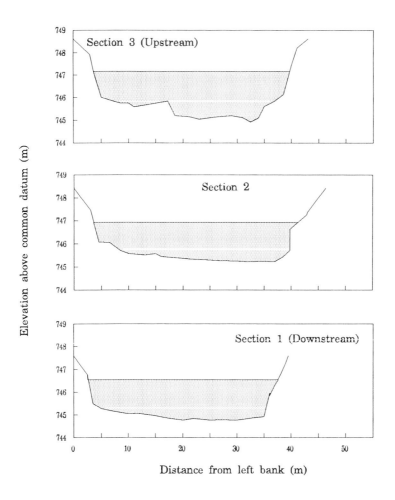

Plan (not to scale) and cross sections, Clarence at Jollies.

n = 0.037

74347: Loganburn at Gorge (Upstream).

Map reference:-	H43:671233 (Metric); S144:627214 (Yard).
Catchment area:-	150 km².
Period of record:-	July 1980 - 1990.
Mean annual flood:-	37.0 m³/s.
Mean flow:-	1.64 m³/s.

Surveyed reach:-

Cross-sections:-	3 along a 50 m reach.
Manning's n range:-	0.021-0.043
Channel description:-	Bed is composed of boulders and large cobbles. Banks are lined with grass.

Bed Surface Material

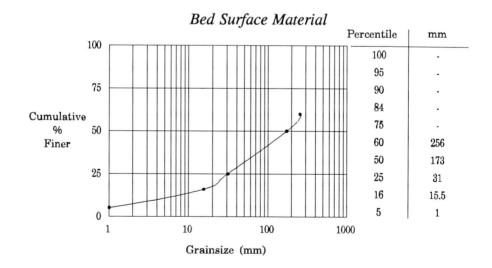

Percentile	mm
100	.
95	.
90	.
84	.
75	.
60	256
50	173
25	31
16	15.5
5	1

Grainsize (mm)

View downstream from top of reach.

View upstream from bottom of reach.

n = 0.037

Hydraulic Properties of Reach

Discharge	Water Surface Slope	Friction Slope	Area	Expansion	Hydraulic Radius	Mean Velocity	Manning n	Chezy C	Error
(m³/s)			(m²)	(%)	(m)	(m/s)			(%)
0.94*	0.00300	0.00130	1.90	-73	0.22	0.70	0.021	36.5	15
2.97*	0.00504	0.00328	4.37	-63	0.44	0.83	0.043	20.0	11
3.68*	0.00448	0.00282	4.87	-57	0.48	0.88	0.041	21.8	10
4.06*	0.00412	0.00210	4.73	-57	0.47	0.99	0.030	29.1	13
4.20*	0.00374	0.00207	5.05	-53	0.50	0.94	0.032	27.3	12
5.82*	0.00456	0.00165	5.33	-56	0.52	1.23	0.022	40.9	20

*   Estimated from rating based on gaugings

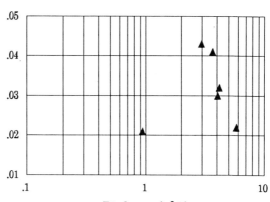

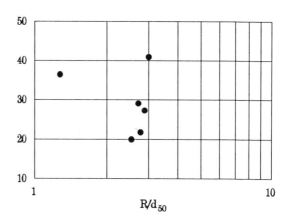

$$n = 0.037$$

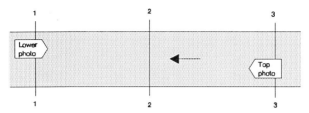

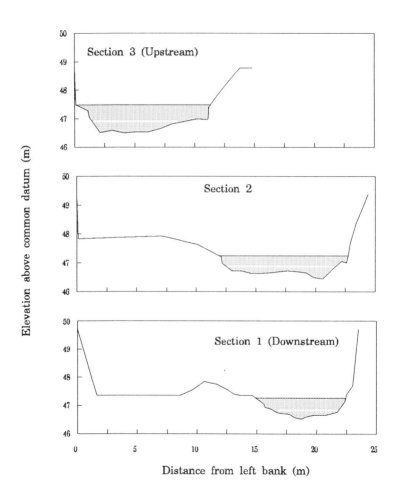

Plan (not to scale) and cross sections, Loganburn at Gorge (Upstream).

137

n = 0.038

91405: Arnold at Lake Brunner.

Map reference:-	K32:844467 (Metric); S051:973727 (Yard).
Catchment area:-	440 km².
Period of record:-	February 1968 - Present.
Mean annual flood:-	193 m³/s.
Mean flow:-	58.2 m³/s.

Surveyed reach:-

Cross-sections:-	5 along a 435 m reach.
Manning's n range:-	0.034-0.040
Channel description:-	Bed material is sand and fine cobbles. Banks are bush covered, with trees overhanging channel and grass under trees. Reach is subject to weed growth.

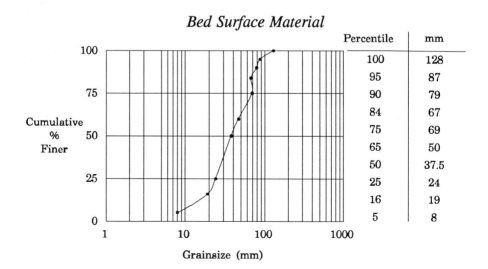

Bed Surface Material

Percentile	mm
100	128
95	87
90	79
84	67
75	69
65	50
50	37.5
25	24
16	19
5	8

View downstream from top cross-section.

View upstream from bottom cross-section.

n = 0.038

Hydraulic Properties of Reach

Discharge	Water Surface Slope	Friction Slope	Area	Expansion	Hydraulic Radius	Mean Velocity	Manning n	Chezy C	Error
(m³/s)			(m²)	(%)	(m)	(m/s)			(%)
24.3	0.00092	0.00093	34.2	31	0.84	0.74	0.039	25.1	8
36.8	0.00098	0.00100	41.7	27	0.98	0.91	0.036	27.7	8
44.4	0.00106	0.00107	47.8	23	1.10	0.95	0.039	26.4	8
72.2	0.00115	0.00116	63.3	19	1.55	1.15	0.040	26.7	8
84.4	0.00109	0.00111	66.5	21	1.44	1.28	0.034	31.1	8
125*	0.00114	0.00117	95.5	17	1.89	1.31	0.040	27.7	8

* Estimated from rating based on gaugings

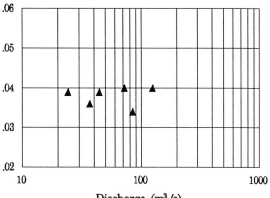

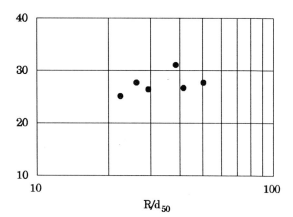

$n = 0.038$

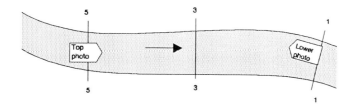

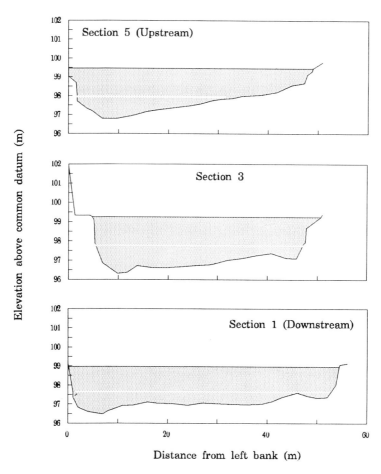

Plan (not to scale) and cross sections, Arnold at Lake Brunner.

n = 0.040 (est.)

32702: Rangitikei at Mangaweka.

Map reference:-	T22:504513 (Metric); N139:288015 (Yard).
Catchment area:-	2787 km².
Period of record:-	December 1953 - 1989.
Mean annual flood:-	739 m³/s.
Mean flow:-	62.9 m³/s.

Surveyed reach:-

Cross-sections:-	5 along a 849 m reach.
Manning's n range:-	0.034-0.048
Channel description:-	Bed comprises gravel and cobbles with scattered small boulders. Banks are composed of Papa mudstone, partly covered with bush, scrub, or grass.

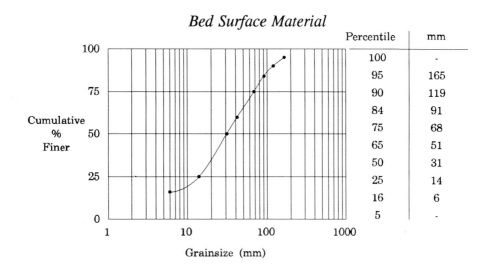

Bed Surface Material

Percentile	mm
100	-
95	165
90	119
84	91
75	68
65	51
50	31
25	14
16	6
5	-

View upstream to top of reach.

View downstream towards bottom of reach.

n = 0.040 (est.)

Hydraulic Properties of Reach

Discharge	Water Surface Slope	Friction Slope	Area	Expansion	Hydraulic Radius	Mean Velocity	Manning n	Chezy C	Error
(m³/s)			(m²)	(%)	(m)	(m/s)			(%)
15.3*	0.00347	0.00343	20.1	-7	0.57	0.82	0.048	18.8	8
21.9*	0.00348	0.00342	23.8	-9	0.62	0.97	0.043	21.3	8
42.5*	0.00352	0.00343	32.8	-15	0.72	1.34	0.034	27.8	8
144*	0.00358	0.00347	80.1	-20	1.12	1.82	0.035	29.1	8
173*	0.00370	0.00356	102	-22	1.28	1.72	0.042	24.9	8
342*	0.00371	0.00349	158	-21	1.83	2.19	0.040	27.5	8
413*	0.00382	0.00356	182	-24	2.04	2.30	0.041	27.2	8
542*	0.00367	0.00340	220	-21	2.34	2.49	0.042	27.7	8

* Estimated from rating based on gaugings

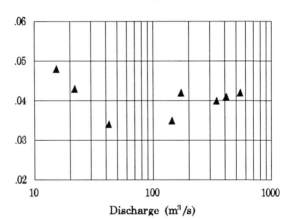

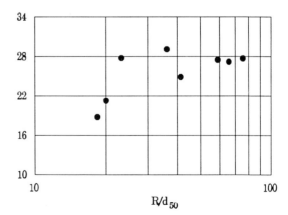

144

$n - 0.010$ (cst.)

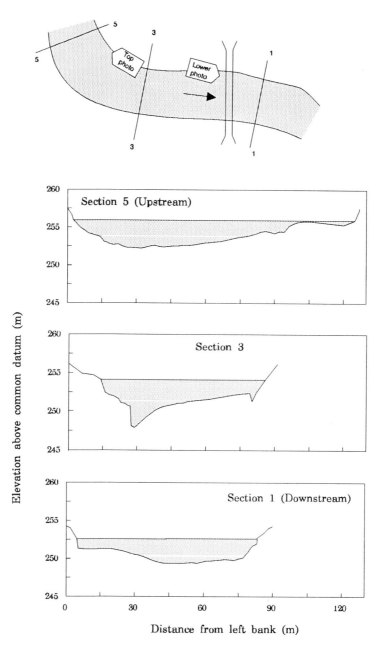

Plan (not to scale) and cross sections, Rangitikei at Mangaweka.

n = 0.04

93208: Buller at Woolfs.

Map reference:-	L29:261297 (Metric); S031:445627 (Yard).
Catchment area:-	4560 km^2.
Period of record:-	October 1963 - Present.
Mean annual flood:-	2900 m^3/s.
Mean flow:-	258 m^3/s.

Surveyed reach:-

Cross-sections:-	5 along a 1208 m reach.
Manning's n range:-	0.030-0.058
Channel description:-	Bed material is mainly silt and small cobbles. Banks are composed of silt and cobbles and are bush covered, with trees overhanging channel.

Bed Surface Material

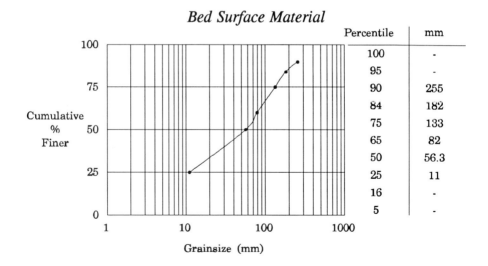

Percentile	mm
100	-
95	-
90	255
84	182
75	133
65	82
50	56.3
25	11
16	-
5	-

View upstream to top cross-section.

View downstream to bottom cross-section.

$n = 0.040$

Hydraulic Properties of Reach

Discharge	Water Surface Slope	Friction Slope	Area	Expansion	Hydraulic Radius	Mean Velocity	Manning n	Chezy C	Error
(m³/s)			(m²)	(%)	(m)	(m/s)			(%)
92.1	0.00060	0.00057	193	-63	1.60	0.53	0.058	18.3	8
124	0.00057	0.00054	189	-59	1.46	0.73	0.038	27.9	8
149	0.00055	0.00051	216	-55	1.74	0.74	0.041	26.1	8
285	0.00074	0.00065	275	-54	2.16	1.11	0.037	30.1	8
573	0.00075	0.00065	381	-42	2.85	1.55	0.032	37.2	8
1079*	0.00086	0.00073	527	-34	3.75	2.10	0.030	41.3	8
2810*	0.00125	0.00098	901	-27	5.64	3.19	0.030	44.4	9

* Estimated from rating based on gaugings

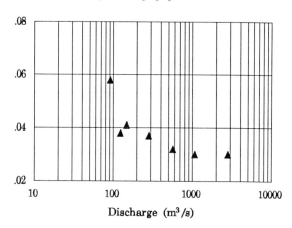

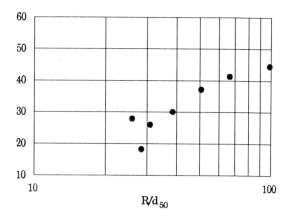

148

$n = 0.040$

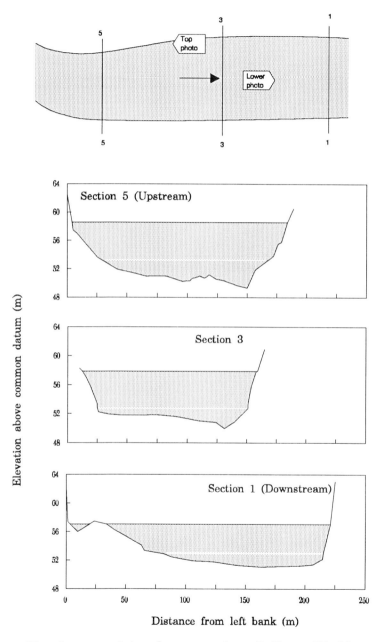

Plan (not to scale) and cross sections, Buller at Woolfs.

n = 0.041

1043428: Tahunatara at Ohakuri Road.

Map reference:-	U16:787140 (Metric); N085:548802 (Yard).
Catchment area:-	210 km².
Period of record:-	April 1964 - December 1988.
Mean annual flood:-	39.5 m³/s.
Mean flow:-	4.77 m³/s.

Surveyed reach:-

Cross-sections:-	5 along a 263 m reach.
Manning's n range:-	0.027-0.049
Channel description:-	Bed is composed of pumice sand. Banks are formed of silt and are grassed down to mean water level.

Bed Surface Material

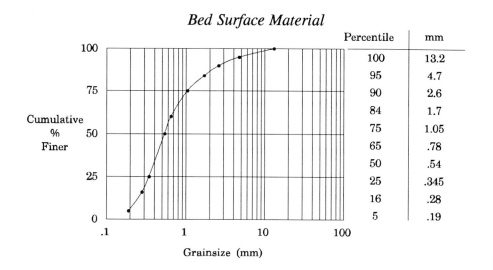

Percentile	mm
100	13.2
95	4.7
90	2.6
84	1.7
75	1.05
65	.78
50	.54
25	.345
16	.28
5	.19

View from middle section looking upstream.

View from middle section looking downstream.

n = 0.041

Hydraulic Properties of Reach

Discharge	Water Surface Slope	Friction Slope	Area	Expansion	Hydraulic Radius	Mean Velocity	Manning n	Chezy C	Error
(m³/s)			(m²)	(%)	(m)	(m/s)			(%)
2.93	0.00018	0.00018	11.8	-2	0.88	0.25	0.049	20.1	10
7.45*	0.00027	0.00026	15.3	-3	1.06	0.49	0.034	29.7	9
9.97	0.00036	0.00035	17.1	-3	1.14	0.59	0.034	29.6	9
15.6*	0.00036	0.00035	20.6	-2	1.30	0.76	0.029	36.0	9
18.1	0.00044	0.00042	22.3	-2	1.36	0.84	0.031	34.2	9
21.8*	0.00051	0.00048	22.3	-4	1.36	0.98	0.027	38.5	10
36.0	0.00063	0.00060	31.9	-1	1.58	1.13	0.029	36.7	10

* Estimated from rating based on gaugings

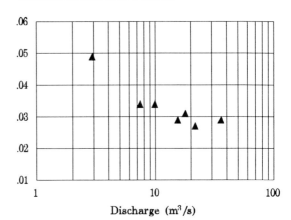

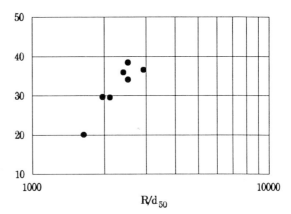

$$n = 0.041$$

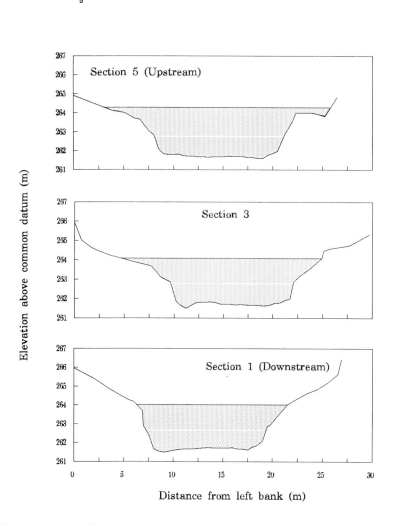

Plan (not to scale) and cross sections, Tahunatara at Ohakuri Road.

n = 0.041

1014641: Ngongotaha at SH5 Bridge.

Map reference:-	U15:900414 (Metric); N076:662105 (Yard).
Catchment area:-	73.3 km².
Period of record:-	May 1975 - December 1988.
Mean annual flood:-	21 m³/s.
Mean flow:-	1.7 m³/s.

Surveyed reach:-

Cross-sections:-	4 along a 115 m reach.
Manning's n range:-	0.044-0.090
Channel description:-	Bed material comprises pumice sand and fine gravel. Banks have willows and other vegetation overhanging the channel.

Bed Surface Material

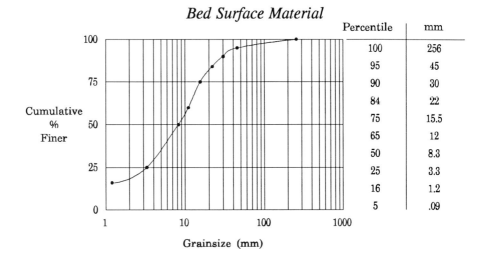

Percentile	mm
100	256
95	45
90	30
84	22
75	15.5
65	12
50	8.3
25	3.3
16	1.2
5	.09

Grainsize (mm)

$n = 0.041$

View downstream from middle of reach.

View upstream from bottom of reach.

n = 0.041

Hydraulic Properties of Reach

Discharge	Water Surface Slope	Friction Slope	Area	Expansion	Hydraulic Radius	Mean Velocity	Manning n	Chezy C	Error
(m³/s)			(m²)	(%)	(m)	(m/s)			(%)
1.89*	0.00080	0.00082	4.39	-19	0.64	0.45	0.044	21.0	9
2.07	0.00095	0.00086	4.69	-18	0.67	0.46	0.045	20.6	9
4.05	0.00098	0.00092	7.33	-13	0.92	0.56	0.048	20.5	9
5.50*	0.00095	0.00089	9.79	-10	1.06	0.57	0.052	19.3	9
5.95*	0.00099	0.00094	10.3	-9	1.08	0.59	0.053	19.2	9
7.19	0.00106	0.00099	12.1	-9	1.15	0.60	0.055	18.6	9
7.79	0.00105	0.00099	13.2	-8	1.17	0.60	0.056	18.2	9
8.98	0.00097	0.00092	15.9	-7	1.24	0.57	0.058	17.7	9
12.0	0.00104	0.00099	20.6	-5	1.42	0.59	0.065	16.2	9
27.7	0.00133	0.00129	44.7	3	2.02	0.63	0.090	12.5	8

* Estimated from rating based on gaugings

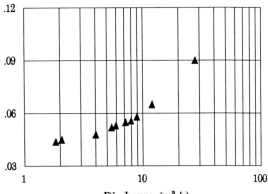

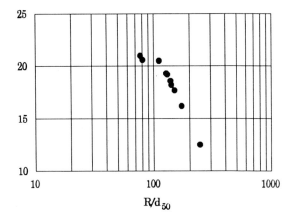

156

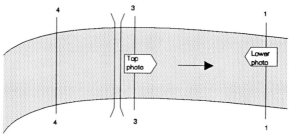

$$n = 0.041$$

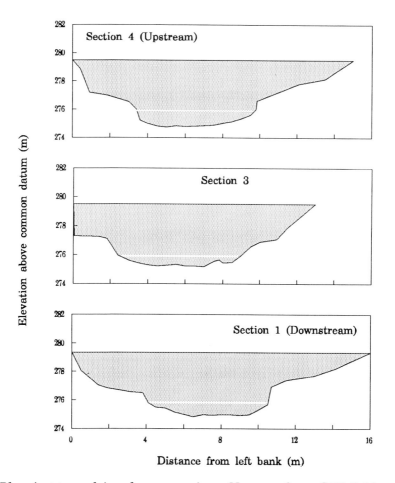

Plan (not to scale) and cross sections, Ngongotaha at SH5 Bridge.

n = 0.041

33301: Wanganui at Paetawa.

Map reference:-	S22:937566 (Metric); N138:667055 (Yard).
Catchment area:-	6640 km².
Period of record:-	July 1957 - 1989.
Mean annual flood:-	2200 m³/s.
Mean flow:-	211 m³/s.

Surveyed reach:-

Cross-sections:-	5 along a 614 m reach.
Manning's n range:-	0.032-0.050
Channel description:-	Bed consists of sand to small cobbles. Right bank is clean; left bank is lined with willows.

Bed Surface Material

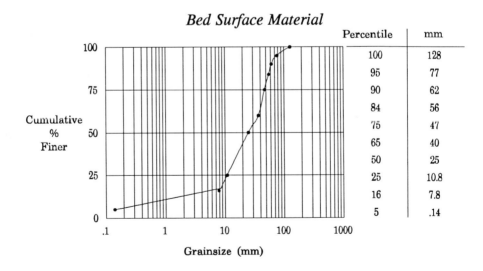

Percentile	mm
100	128
95	77
90	62
84	56
75	47
65	40
50	25
25	10.8
16	7.8
5	.14

View upstream at top cross-section.

View downstream at bottom cross-section.

n = 0.041

Hydraulic Properties of Reach

Discharge	Water Surface Slope	Friction Slope	Area	Expansion	Hydraulic Radius	Mean Velocity	Manning n	Chezy C	Error
(m³/s)			(m²)	(%)	(m)	(m/s)			(%)
32.6*	0.00010	0.00010	128	48	1.54	0.26	0.048	22.0	9
45.9*	0.00012	0.00012	149	39	1.76	0.31	0.050	22.0	9
130*	0.00029	0.00029	203	22	2.32	0.64	0.046	25.1	8
381*	0.00040	0.00041	320	11	3.57	1.19	0.039	31.9	11
962*	0.00030	0.00030	565	5	5.52	1.70	0.032	41.7	9
1190*	0.00029	0.00029	658	4	6.14	1.81	0.032	42.9	14
1810*	0.00028	0.00028	915	2	7.38	1.98	0.032	43.6	15
2130*	0.00029	0.00028	1050	1	7.97	2.03	0.033	42.8	18
2960*	0.00026	0.00026	1420	3	9.17	2.08	0.034	42.3	21

* Estimated from rating based on gaugings

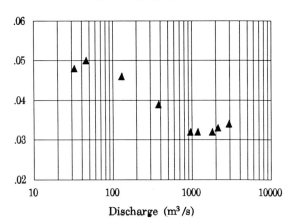

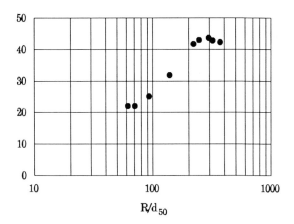

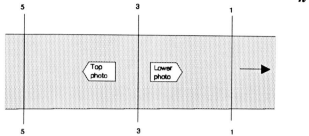

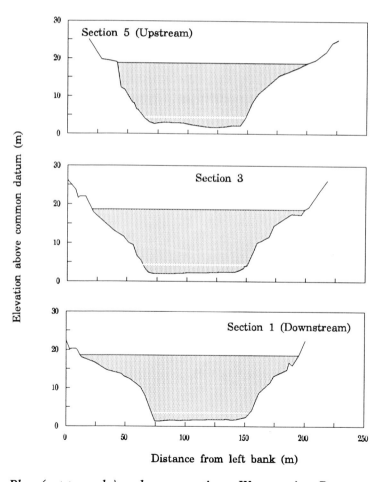

Plan (not to scale) and cross sections, Wanganui at Paetawa.

n = 0.042

21409: Waiau at Otoi.

Map reference:-	W19:620427 (Metric); N105:480048 (Yard).
Catchment area:-	513 km².
Period of record:-	August 1968 - December 1989.
Mean annual flood:-	331 m³/s.
Mean flow:-	20.5 m³/s.

Surveyed reach:-

Cross-sections:-	3 along a 214 m reach.
Manning's n range:-	0.040-0.044
Channel description:-	Bed consists of gravel and boulders. Right bank has steep mudstone banks with some overhanging vegetation; left bank is grassed with trees.

View downstream to bridge from cross-section 3.

View upstream to bridge from cross-section 1.

n = 0.042

Hydraulic Properties of Reach

Discharge	Water Surface Slope	Friction Slope	Area	Expansion	Hydraulic Radius	Mean Velocity	Manning n	Chezy C	Error
(m³/s)			(m²)	(%)	(m)	(m/s)			(%)
14.7*	0.00254	0.00253	18.4	1	0.54	0.80	0.042	21.5	8
15.8*	0.00245	0.00244	19.1	1	0.56	0.83	0.040	22.4	8
16.5*	0.00249	0.00247	19.5	1	0.56	0.84	0.040	22.6	8
17.7*	0.00259	0.00258	21.0	1	0.60	0.84	0.042	21.7	8
18.2*	0.00264	0.00263	21.1	0	0.60	0.86	0.041	22.0	8
21.9*	0.00280	0.00278	23.4	0	0.65	0.93	0.042	22.1	8
27.5*	0.00301	0.00298	26.7	0	0.73	1.03	0.043	22.1	8
35.7*	0.00326	0.00324	31.4	0	0.84	1.14	0.044	21.8	8
51.6*	0.00344	0.00344	38.5	4	1.00	1.34	0.044	22.7	8
58.3*	0.00354	0.00354	41.0	5	1.05	1.42	0.043	23.1	8

* Estimated from rating based on gaugings

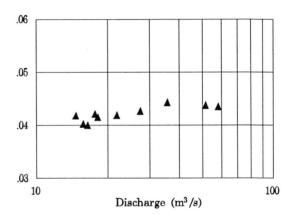

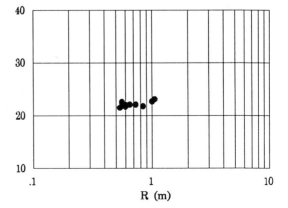

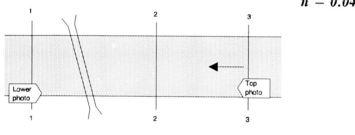

$n - 0.042$

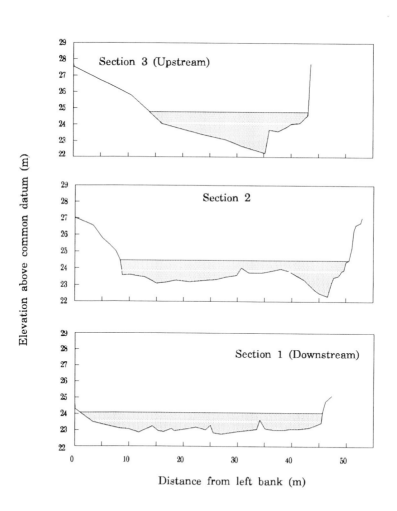

Plan (not to scale) and cross sections, Waiau at Otoi.

n = 0.043

15412: Rangitaiki at Te Teko.

Map reference:-	V15:436444 (Metric); N077:248155 (Yard).
Catchment area:-	2893 km².
Period of record:-	September 1952 - January 1989.
Mean annual flood:-	263 m³/s.
Mean flow:-	71.4 m³/s.

Surveyed reach:-

Cross-sections:-	4 along a 425 m reach.
Manning's n range:-	0.042-0.050
Channel description:-	Bed material consists of sand with some gravel. Bank material is mainly silt and sand. Willows overhang the banks along more than half the reach. Lower banks are kept clear of vegetation by water level fluctuations associated with hydro operations.

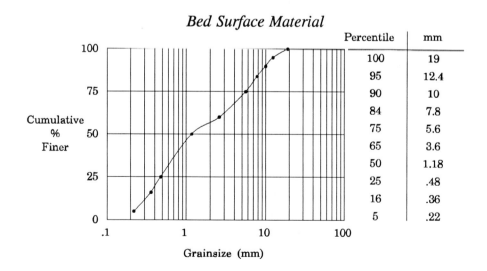

Bed Surface Material

Percentile	mm
100	19
95	12.4
90	10
84	7.8
75	5.6
65	3.6
50	1.18
25	.48
16	.36
5	.22

View downstream from bottom of reach.

View upstream from bottom of reach.

$n = 0.043$

Hydraulic Properties of Reach

Discharge	Water Surface Slope	Friction Slope	Area	Expansion	Hydraulic Radius	Mean Velocity	Manning n	Chezy C	Error
(m³/s)			(m²)	(%)	(m)	(m/s)			(%)
47.5*	0.00062	0.00057	68.6	-29	1.71	0.70	0.050	22.0	8
53.0*	0.00054	0.00050	72.3	-25	1.78	0.74	0.045	24.5	8
74.0	0.00050	0.00047	88.8	-14	2.07	0.84	0.042	26.7	8
98.0*	0.00055	0.00053	110	-4	2.35	0.89	0.046	25.3	8
107	0.00050	0.00049	118	0	2.46	0.91	0.044	26.3	9
120*	0.00047	0.00047	129	5	2.61	0.93	0.044	27.0	9
144	0.00046	0.00046	150	8	2.73	0.96	0.043	27.5	9

* Estimated from rating based on gaugings

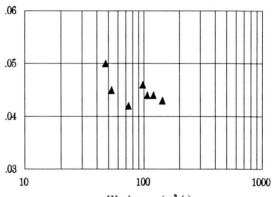

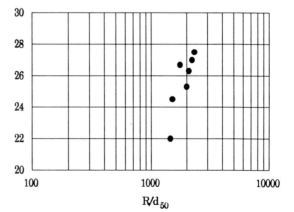

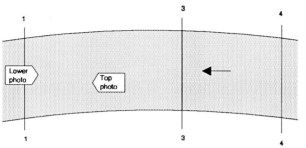

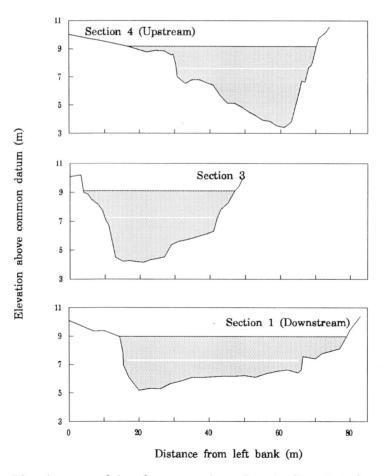

Plan (not to scale) and cross sections, Rangitaiki at Te Teko.

n = 0.044

29809: Hutt at Taita Gorge.

Map reference:-	R27:764034 (Metric); N160:525375 (Yard).
Catchment area:-	555 km².
Period of record:-	March 1979 - January 1990.
Mean annual flood:-	869 m³/s.
Mean flow:-	25.2 m³/s.

Surveyed reach:-

Cross-sections:-	3 along a 193 m reach.
Manning's n range:-	0.031-0.044
Channel description:-	Bed consists of cobbles and gravel. Both banks are lined with scrub and overhanging trees.

Bed Surface Material

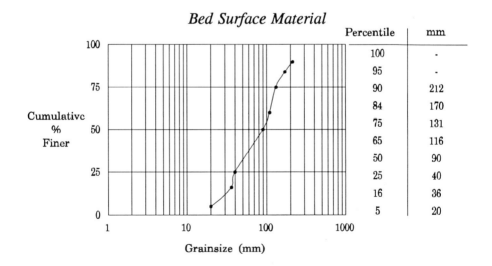

Percentile	mm
100	-
95	-
90	212
84	170
75	131
65	116
50	90
25	40
16	36
5	20

$n = 0.044$

View downstream from top cross-section.

View upstream from bottom cross-section.

n = 0.044

Hydraulic Properties of Reach

Discharge	Water Surface Slope	Friction Slope	Area	Expansion	Hydraulic Radius	Mean Velocity	Manning n	Chezy C	Error
(m³/s)			(m²)	(%)	(m)	(m/s)			(%)
23.8*	0.00174	0.00173	33.0	2	0.63	0.72	0.044	21.1	8
59.4	0.00182	0.00183	52.3	6	0.97	1.14	0.038	26.5	8
79.0	0.00189	0.00191	61.6	7	1.12	1.28	0.037	27.3	9
93.0	0.00195	0.00197	69.2	7	1.24	1.34	0.039	26.7	9
137	0.00184	0.00190	95.8	11	1.65	1.43	0.043	25.4	10
298	0.00199	0.00212	124	9	2.06	2.40	0.031	36.4	9

* Estimated from rating based on gaugings

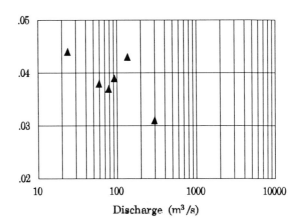

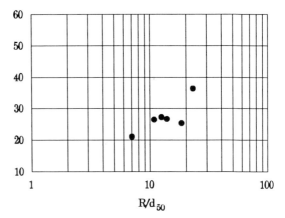

$n = 0.044$

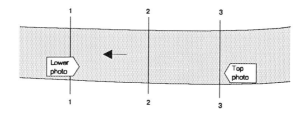

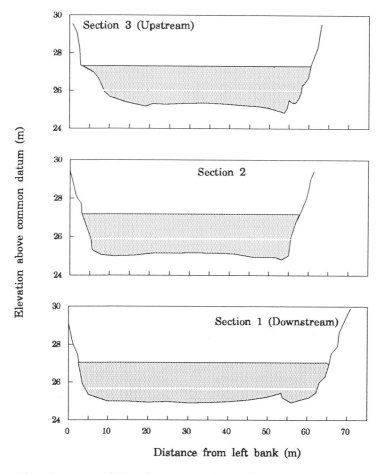

Plan (not to scale) and cross sections, Hutt at Taita Gorge.

n = 0.045 (est.)

45703: Hoteo at Gubbs.

Map reference:-	Q09:460340 (Metric); N033:029167 (Yard).
Catchment area:-	268 km².
Period of record:-	August 1977 - Present.
Mean annual flood:-	146 m³/s.
Mean flow:-	7.16 m³/s.

Surveyed reach:-

Cross-sections:-	3 along a 236 m reach.
Manning's n range:-	0.032-0.061
Channel description:-	Bed is mostly silt and clay/papa with numerous old tree roots and trunks protruding. The silt banks are continually slipping into the river.

$n = 0.045$ (est.)

View from cross-section 2 looking upstream.

View from cross-section 2 looking downstream.

n = 0.045 (est.)

Hydraulic Properties of Reach

Discharge	Water Surface Slope	Friction Slope	Area	Expansion	Hydraulic Radius	Mean Velocity	Manning n	Chezy C	Error
(m³/s)			(m²)	(%)	(m)	(m/s)			(%)
24.1*	0.00099	0.00102	30.2	10	1.82	0.80	0.060	18.5	10
27.9*	0.00111	0.00111	33.1	3	1.88	0.84	0.061	18.2	8
39.4*	0.00094	0.00094	40.5	2	2.01	0.97	0.051	22.1	9
39.8*	0.00111	0.00113	41.6	4	2.03	0.96	0.056	20.0	11
52.3*	0.00078	0.00079	48.4	2	2.16	1.08	0.044	26.0	9
54.3*	0.00076	0.00076	50.9	1	2.21	1.07	0.044	25.8	9
72.1*	0.00053	0.00053	60.6	1	2.43	1.19	0.035	33.0	12
99.2*	0.00043	0.00042	80.2	0	2.62	1.24	0.032	37.0	13
149*	0.00054	0.00051	120	-3	2.92	1.24	0.037	32.0	12
156*	0.00064	0.00061	129	-4	3.03	1.21	0.043	28.1	11

* Estimated from rating based on gaugings

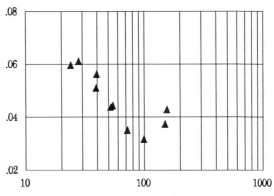

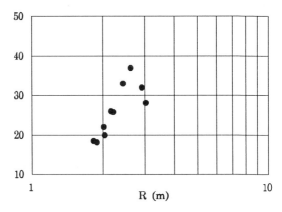

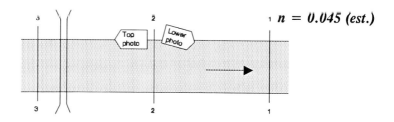

$n = 0.045$ (est.)

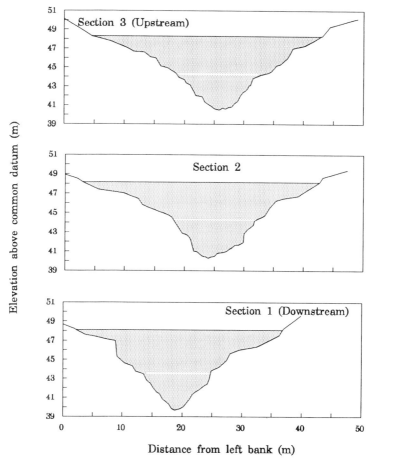

Section 3 (Upstream)

Section 2

Section 1 (Downstream)

Elevation above common datum (m)

Distance from left bank (m)

Plan (not to scale) and cross sections, Hoteo at Gubbs.

n = 0.046

43433: Waipa at Whatawhata.

Map reference:-	S14:997760 (Metric); N065:664456 (Yard).
Catchment area:-	2826 km².
Period of record:-	September 1969 - Present.
Mean annual flood:-	388 m³/s.
Mean flow:-	84.6 m³/s.

Surveyed reach:-

Cross-sections:-	3 along a 171 m reach.
Manning's n range:-	0.034-0.050
Channel description:-	Bed consists mainly of sand and gravel, with a little mud and some large pieces of coal and pumice. Banks are lined with willows and patches of blackberry.

Bed Surface Material

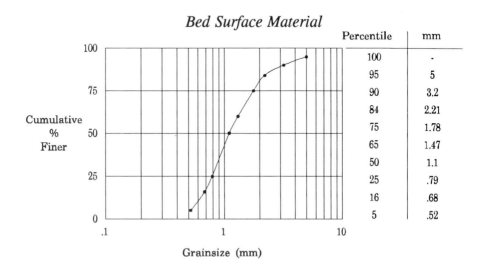

Percentile	mm
100	.
95	5
90	3.2
84	2.21
75	1.78
65	1.47
50	1.1
25	.79
16	.68
5	.52

View downstream from cross-section 3.

View upstream from cross-section 2.

n = 0.046

Hydraulic Properties of Reach

Discharge	Water Surface Slope	Friction Slope	Area	Expansion	Hydraulic Radius	Mean Velocity	Manning n	Chezy C	Error
(m³/s)			(m²)	(%)	(m)	(m/s)			(%)
63.0*	0.00018	0.00017	125	0	2.66	0.50	0.050	23.3	26
129*	0.00012	0.00012	195	1	3.78	0.66	0.040	31.4	31
317*	0.00012	0.00012	356	2	5.25	0.89	0.038	35.1	30
434*	0.00013	0.00013	447	2	4.89	0.97	0.034	38.1	22
623*	0.00019	0.00017	632	0	4.96	0.99	0.038	33.8	14

* Estimated from rating based on gaugings

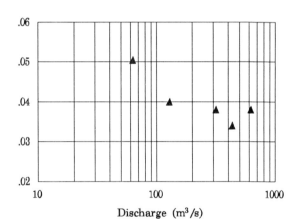

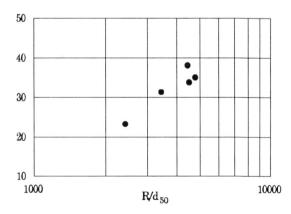

$n = 0.046$

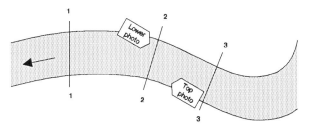

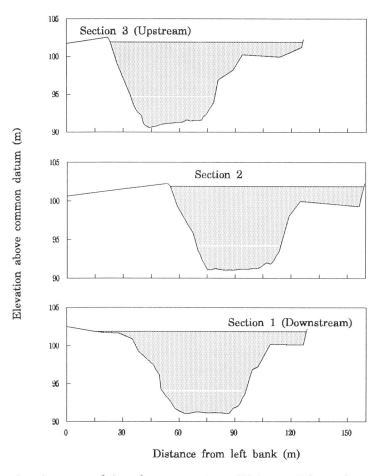

Plan (not to scale) and cross sections, Waipa at Whatawhata.

n = 0.046

40708: Mokau at Totoro Bridge.

Map reference:-	R17:759908 (Metric); N091:431517 (Yard).
Catchment area:-	1046 km².
Period of record:-	April 1979 - January 1990.
Mean annual flood:-	244 m³/s.
Mean flow:-	34.1 m³/s.

Surveyed reach:-	
Cross-sections:-	3 along a 195 m reach.
Manning's n range:-	0.040-0.063
Channel description:-	Bed is generally rock, with some small sections of silt, sand, or gravel beach. Banks are mainly clay and are lined with overhanging willows.

View downstream at cross-section 3.

View upstream at cross-section 1.

n = 0.046

Hydraulic Properties of Reach

Discharge	Water Surface Slope	Friction Slope	Area	Expansion	Hydraulic Radius	Mean Velocity	Manning n	Chezy C	Error
(m³/s)			(m²)	(%)	(m)	(m/s)			(%)
8.86*	0.00016	0.00016	30.2	-6	1.12	0.30	0.044	23.1	13
98.2*	0.00145	0.00122	68.5	-21	2.15	1.45	0.040	28.6	10
195*	0.00175	0.00151	126	-18	3.11	1.57	0.053	23.0	9
240*	0.00131	0.00111	164	-17	3.52	1.47	0.052	23.7	10
255*	0.00180	0.00152	156	-18	3.44	1.65	0.054	22.9	10
271*	0.00122	0.00101	185	-18	3.77	1.47	0.052	24.0	10
327*	0.00144	0.00118	230	-22	4.28	1.44	0.063	20.3	10
349§	0.00184	0.00150	218	-23	4.21	1.62	0.063	20.3	10

* Estimated from rating based on gaugings   § Estimated from theoretical rating for structure

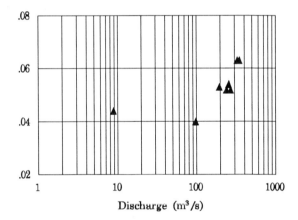

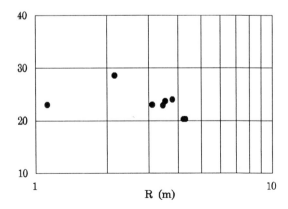

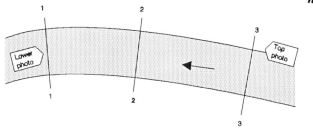

$n - 0.046$

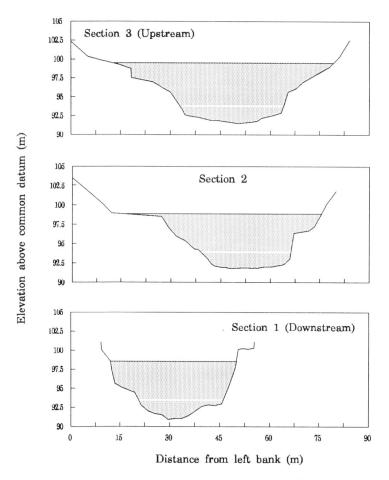

Plan (not to scale) and cross sections, Mokau at Totoro Bridge.

n = 0.047

71129: Forks at Balmoral.

Map reference:-	I37:014892 (Metric); S089:034021 (Yard).
Catchment area:-	98 km².
Period of record:-	July 1962 - Present.
Mean annual flood:-	22 m³/s.
Mean flow:-	3.19 m³/s.

Surveyed reach:-

Cross-sections:-	5 along a 128 m reach.
Manning's n range:-	0.039-0.048
Channel description:-	Bed consists of cobbles and small boulders. Banks are covered with short tussock.

Bed Surface Material

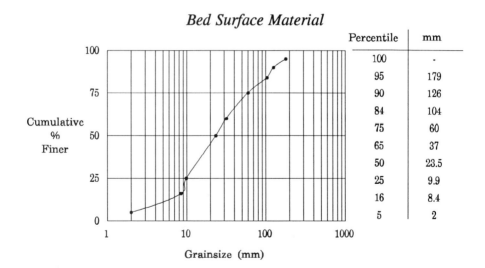

Percentile	mm
100	-
95	179
90	126
84	104
75	60
65	37
50	23.5
25	9.9
16	8.4
5	2

$n = 0.04/$

View upstream towards top of reach.

View downstream towards bottom of reach.

_n = 0.047_

Hydraulic Properties of Reach

Discharge	Water Surface Slope	Friction Slope	Area	Expansion	Hydraulic Radius	Mean Velocity	Manning n	Chezy C	Error
(m³/s)			(m²)	(%)	(m)	(m/s)			(%)
1.97	0.00538	0.00528	3.04	4	0.28	0.68	0.048	16.7	8
3.92	0.00548	0.00535	4.40	6	0.39	0.91	0.046	18.8	8
5.42*	0.00532	0.00520	5.22	12	0.46	1.06	0.043	20.7	8
6.92*	0.00534	0.00521	5.83	15	0.51	1.21	0.040	22.5	8
7.84	0.00524	0.00511	6.39	17	0.55	1.26	0.039	22.1	8
8.00*	0.00524	0.00516	6.53	19	0.56	1.24	0.041	22.4	8

* Estimated from rating based on gaugings

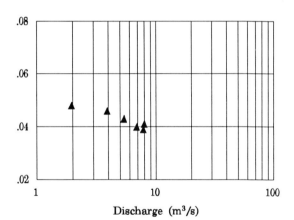

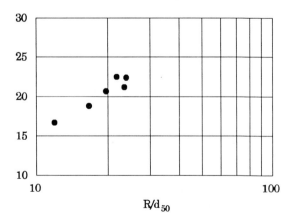

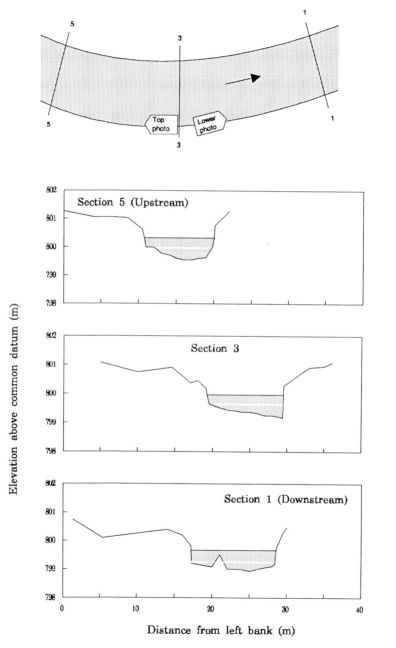

$n = 0.047$

Plan (not to scale) and cross sections, Forks at Balmoral.

n = 0.047

93213: Gowan at Lake Rotorua.

Map reference:-	M29:763348 (Metric); S033:995673 (Yard).
Catchment area:-	368 km².
Period of record:-	March 1934 - January 1990.
Mean annual flood:-	85 m³/s.
Mean flow:-	26.4 m³/s.

Surveyed reach:-

Cross-sections:-	4 along a 231 m reach.
Manning's n range:-	0.040-0.046
Channel description:-	Bed consists predominantly of boulders with some cobbles. Both banks are tree-lined, with trees overhanging river; left bank is partly grassed.

Bed Surface Material

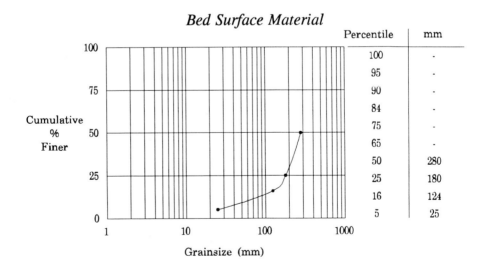

Percentile	mm
100	-
95	-
90	-
84	-
75	-
65	-
50	280
25	180
16	124
5	25

$$n = 0.047$$

View upstream towards top cross-section.

View downstream towards bottom cross-section.

n = 0.047

Hydraulic Properties of Reach

Discharge	Water Surface Slope	Friction Slope	Area	Expansion	Hydraulic Radius	Mean Velocity	Manning n	Chezy C	Error
(m³/s)			(m²)	(%)	(m)	(m/s)			(%)
24.6*	0.00302	0.00305	24.4	27	0.80	1.04	0.046	20.7	8
42.3*	0.00316	0.00319	32.6	24	0.99	1.33	0.044	22.9	8
64.4*	0.00320	0.00325	41.0	23	1.16	1.60	0.041	25.1	8
79.3*	0.00335	0.00338	45.9	19	1.26	1.76	0.040	25.9	9

* Estimated from rating based on gaugings

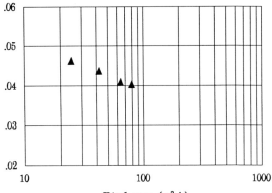

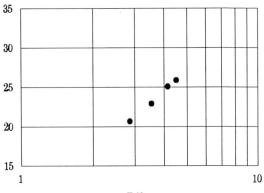

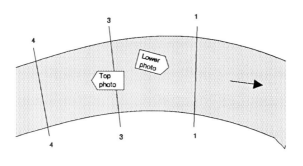

$$n = 0.047$$

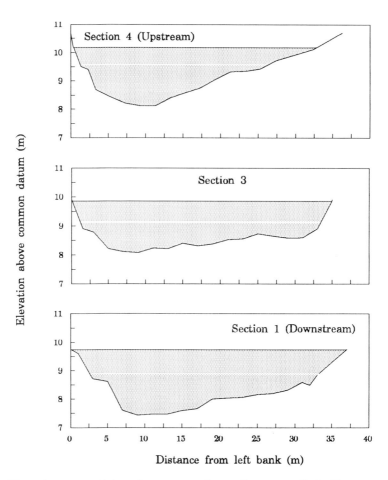

Plan (not to scale) and cross sections, Gowan at Lake Rotorua.

n = 0.049

1043459: Tongariro at Turangi.

Map reference:-	T19:537417 (Metric); N102:297004 (Yard).
Catchment area:-	772 km².
Period of record:-	August 1948 - Present.
Mean annual flood:-	426 m³/s.
Mean flow:-	42.2 m³/s.

Surveyed reach:-

Cross-sections:-	3 along a 325 m reach.
Manning's n range:-	0.028-0.056
Channel description:-	Bed material ranges from sand to large boulders. Banks are lined with trees and bushes.
Note:-	This reach steepens considerably towards the downstream cross-section as the channel changes from pool to riffle morphology. The Manning's n value computed for the pool reach spanned by sections 2 and 3 ranged from 0.014 ($\pm$ 27%) at 27.7 m³/s to 0.018 ($\pm$ 25%) at 161 m³/s.

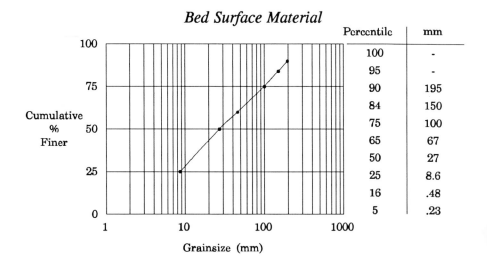

Bed Surface Material

Percentile	mm
100	-
95	-
90	195
84	150
75	100
65	67
50	27
25	8.6
16	.48
5	.23

View from middle cross-section looking upstream.

View from middle cross-section looking downstream.

_n = 0.049_

Hydraulic Properties of Reach[+]

Discharge	Water Surface Slope	Friction Slope	Area	Expansion	Hydraulic Radius	Mean Velocity	Manning n	Chezy C	Error
(m³/s)			(m²)	(%)	(m)	(m/s)			(%)
27.6	0.00261	0.00252	33.9	-22	0.83	0.85	0.056	17.2	8
27.7	0.00248	0.00240	33.6	-18	0.83	0.86	0.055	17.7	8
28.2	0.00259	0.00249	33.5	-22	0.83	0.88	0.054	17.8	8
31.2*	0.00271	0.00259	34.5	-25	0.84	0.94	0.051	18.8	8
46.7	0.00279	0.00261	40.4	-21	0.94	1.21	0.044	22.4	8
161	0.00193	0.00217	76.3	36	1.40	2.16	0.028	37.9	10

* Estimated from rating based on gaugings   + See note, page 194

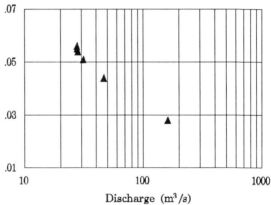

Manning n

Discharge (m³/s)

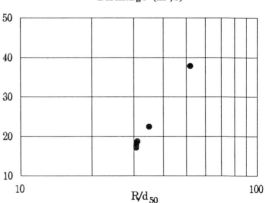

Chezy C

R/d$_{50}$

n = 0.049

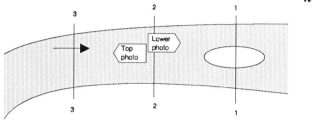

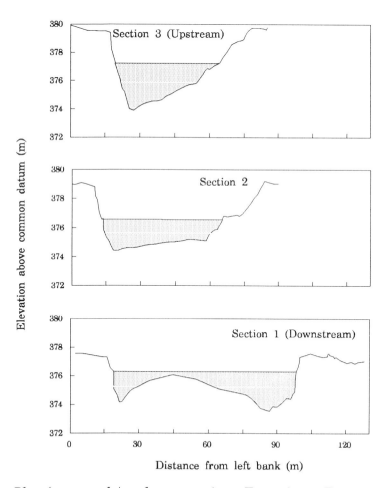

Plan (not to scale) and cross sections, Tongariro at Turangi.

n = 0.049

71135: Jollie at Mount Cook Station.

Map reference:-	H37:835019 (Metric); S089:841164 (Yard).
Catchment area:-	139 km².
Period of record:-	December 1964 - Present.
Mean annual flood:-	70 m³/s.
Mean flow:-	7.97 m³/s.

Surveyed reach:-

Cross-sections:-	5 along a 128 m reach.
Manning's n range:-	0.039-0.047
Channel description:-	Bed consists of large cobbles and boulders. Banks are lined mostly with brush and some willows.

Bed Surface Material

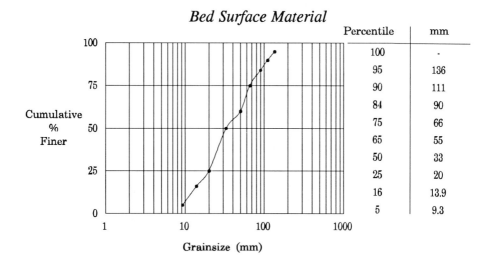

Percentile	mm
100	-
95	136
90	111
84	90
75	66
65	55
50	33
25	20
16	13.9
5	9.3

n = 0.049

View upstream from middle section.

View downstream towards bottom cross-section.

n = 0.049

Hydraulic Properties of Reach

Discharge	Water Surface Slope	Friction Slope	Area	Expansion	Hydraulic Radius	Mean Velocity	Manning n	Chezy C	Error
(m³/s)			(m²)	(%)	(m)	(m/s)			(%)
9.25	0.00812	0.00846	8.60	135	0.44	1.16	0.046	19.0	8
16.1*	0.00881	0.00906	11.9	77	0.58	1.42	0.047	19.6	8
17.7*	0.00796	0.00855	11.2	127	0.55	1.70	0.037	24.4	8
18.6	0.00780	0.00842	13.2	133	0.62	1.53	0.045	20.7	8
31.3	0.00945	0.00983	15.8	54	0.73	2.05	0.039	24.5	8

* Estimated from rating based on gaugings

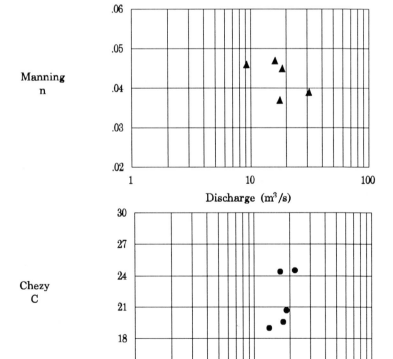

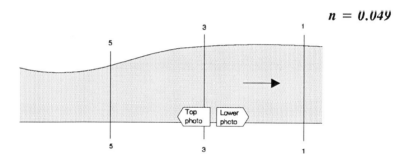

$n = 0.049$

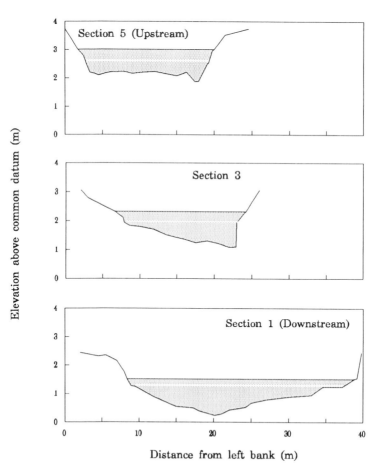

Plan (not to scale) and cross sections, Jollie at Mount Cook Station.

$n = 0.051$

43435: Waipapa at Ngaroma Rd.

Map reference:-	T16:425166 (Metric); N084:151820 (Yard).
Catchment area:-	137 km².
Period of record:-	April 1964 - December 1988.
Mean annual flood:-	50 m³/s.
Mean flow:-	5.69 m³/s.

Surveyed reach:-

Cross-sections:-	4 along a 135 m reach.
Manning's n range:-	0.027-0.055
Channel description:-	Bed material is bedrock along the entire reach. Vegetation overhangs both banks.

View upstream from middle of reach.

View downstream from middle of reach.

n = 0.051

Hydraulic Properties of Reach

Discharge	Water Surface Slope	Friction Slope	Area	Expansion	Hydraulic Radius	Mean Velocity	Manning n	Chezy C	Error
(m³/s)			(m²)	(%)	(m)	(m/s)			(%)
3.50*	0.00911	0.00906	6.31	24	0.33	0.60	0.055	14.3	8
12.5	0.00727	0.00716	12.1	37	0.55	1.11	0.040	21.8	8
22.9	0.00734	0.00710	15.3	22	0.67	1.58	0.034	21.2	8
31.4*	0.00729	0.00692	17.5	18	0.74	1.89	0.030	30.6	8
38.5*	0.00686	0.00636	18.9	21	0.79	2.14	0.027	35.1	9
57.4*	0.00700	0.00632	23.9	11	0.96	2.50	0.027	36.4	9

* Estimated from rating based on gaugings

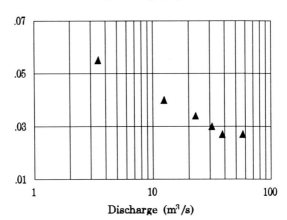

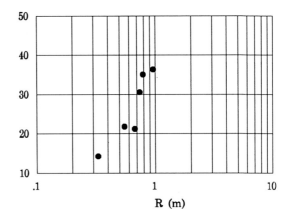

$n = 0.051$

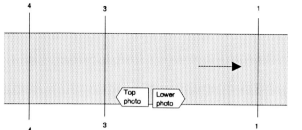

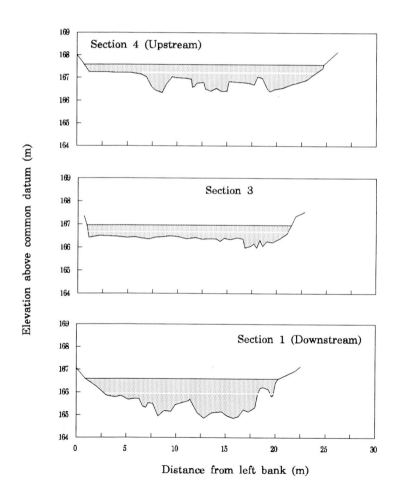

Plan (not to scale) and cross sections, Waipapa at Ngaroma Rd.

n = 0.052

31807: Otaki at Pukehinau.

Map reference:-	S25:955402 (Metric); N157:723783 (Yard).
Catchment area:-	306 km².
Period of record:-	July 1980 - December 1988.
Mean annual flood:-	891 m³/s.
Mean flow:-	31.5 m³/s.

Surveyed reach:-

Cross-sections:-	3 along a 100 m reach.
Manning's n range:-	0.028-0.073
Channel description:-	Bed consists of cobbles and gravel with some sand. Both banks are sheer cliffs, sparsely vegetated.

Bed Surface Material

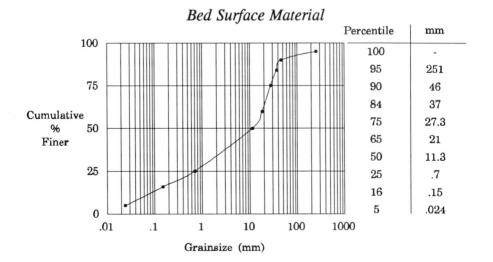

Percentile	mm
100	-
95	251
90	46
84	37
75	27.3
65	21
50	11.3
25	.7
16	.15
5	.024

View downstream from top cross-section.

View upstream from bottom cross-section.

_n = 0.052_

Hydraulic Properties of Reach

Discharge	Water Surface Slope	Friction Slope	Area	Expansion	Hydraulic Radius	Mean Velocity	Manning n	Chezy C	Error
(m³/s)			(m²)	(%)	(m)	(m/s)			(%)
44.5*	0.00020	0.00020	107	5	2.29	0.42	0.060	19.4	18
65.0	0.00020	0.00021	116	7	2.42	0.56	0.046	25.0	19
75.0	0.00050	0.00051	128	6	2.58	0.59	0.073	16.2	11
84.0	0.00018	0.00018	132	3	2.64	0.64	0.041	29.1	23
123	0.00014	0.00016	136	8	2.70	0.90	0.028	42.9	37

* Estimated from rating based on gaugings

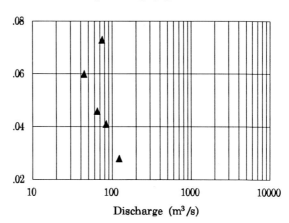

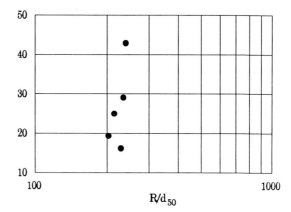

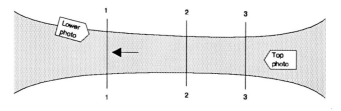

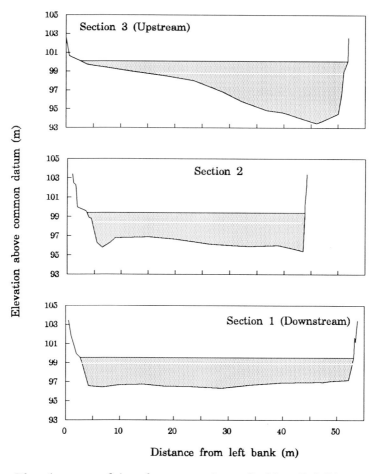

Plan (not to scale) and cross sections, Otaki at Pukehinau.

n = 0.054

57014: Stanley Brook at Barkers.

Map reference:-	N27:949877 (Metric); S019:208248 (Yard).
Catchment area:-	81.6 km².
Period of record:-	November 1969 - Present.
Mean annual flood:-	64.2 m³/s.
Mean flow:-	1.29 m³/s.

Surveyed reach:-

Cross-sections:-	5 along a 191 m reach.
Manning's n range:-	0.026-0.074
Channel description:-	Bed consists primarily of gravel and small cobbles with some bedrock at upper end of reach. Banks are lined with willows and some dense blackberry and scrub. Trees overhang channel on both sides.

Bed Surface Material

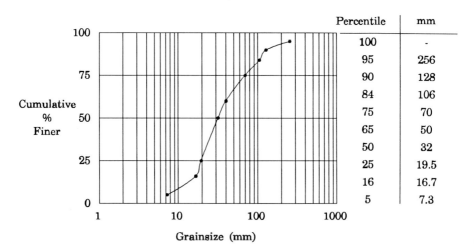

Percentile	mm
100	-
95	256
90	128
84	106
75	70
65	50
50	32
25	19.5
16	16.7
5	7.3

View upstream towards top cross-section.

View upstream to bottom cross-section.

n = 0.054

Hydraulic Properties of Reach

Discharge	Water Surface Slope	Friction Slope	Area	Expansion	Hydraulic Radius	Mean Velocity	Manning n	Chezy C	Error
(m³/s)			(m²)	(%)	(m)	(m/s)			(%)
0.53*	0.00491	0.00491	1.68	31	0.19	0.32	0.074	10.2	8
1.66*	0.00573	0.00425	2.74	-23	0.27	0.62	0.048	16.4	8
2.84*	0.00582	0.00573	3.38	-19	0.32	0.84	0.041	20.1	8
9.64*	0.00586	0.00572	6.01	2	0.50	1.60	0.030	29.9	8
24.0*	0.00659	0.00617	9.89	-5	0.72	2.43	0.026	36.4	9
36.9*	0.00643	0.00605	14.2	8	0.94	2.61	0.028	34.9	9

* Estimated from rating based on gaugings

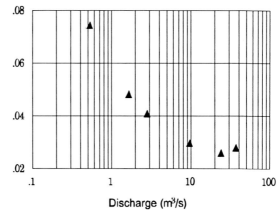

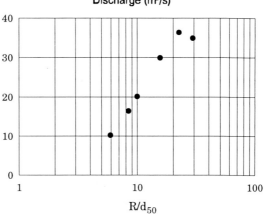

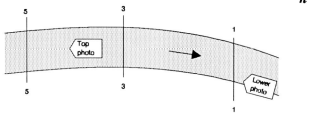

$n - 0.054$

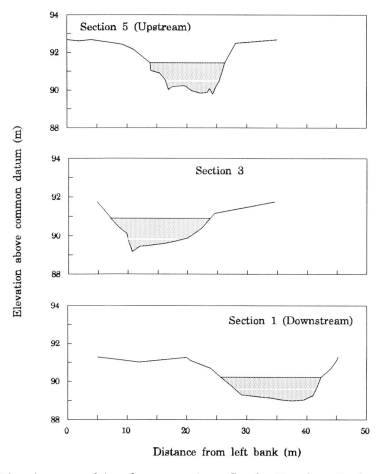

Plan (not to scale) and cross sections, Stanley Brook at Barkers.

n = 0.055 (est.)

25902: Whareama at Waiteko.

Map reference:-	T26:660231 (Metric); N159:499618 (Yard).
Catchment area:-	398 km².
Period of record:-	April 1970 - May 1989.
Mean annual flood:-	308 m³/s.
Mean flow:-	6.56 m³/s.

Surveyed reach:-	
Cross-sections:-	2 along a 109 m reach.
Manning's n range:-	0.048-0.073
Channel description:-	Bed consists of silt with much plant debris lining the bottom of the channel. Banks are grassed and subject to slope failure.

Bed Surface Material

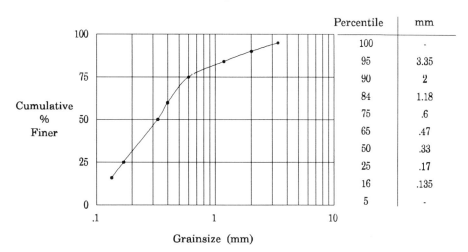

Percentile	mm
100	-
95	3.35
90	2
84	1.18
75	.6
65	.47
50	.33
25	.17
16	.135
5	-

Grainsize (mm)

View downstream from top cross-section.

View upstream from bottom cross-section.

n = 0.055 (est.)

Hydraulic Properties of Reach

Discharge	Water Surface Slope	Friction Slope	Area	Expansion	Hydraulic Radius	Mean Velocity	Manning n	Chezy C	Error
(m³/s)			(m²)	(%)	(m)	(m/s)			(%)
23.1	0.00081	0.00094	35.9	47	2.01	0.68	0.073	15.4	16
26.6	0.00056	0.00065	37.5	25	2.04	0.72	0.057	19.7	25
30.0	0.00057	0.00068	41.7	31	2.14	0.73	0.060	19.0	26
36.0	0.00044	0.00055	48.2	30	2.29	0.76	0.054	21.2	11
200	0.00056	0.00085	139	20	3.67	1.44	0.048	25.9	15
220	0.00056	0.00079	159	17	3.88	1.39	0.050	25.1	13
289*	0.00176	0.00204	170	14	3.96	1.70	0.067	18.9	11

* Estimated from rating based on gaugings

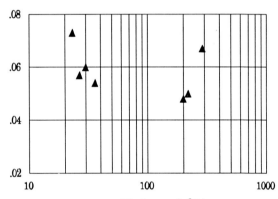

Manning n vs. Discharge (m³/s)

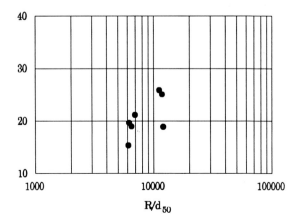

Chezy C vs. R/d_{50}

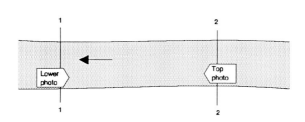

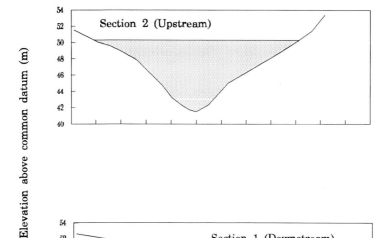

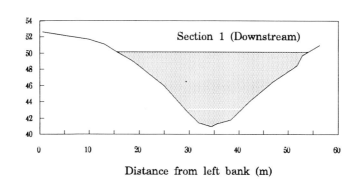

Plan (not to scale) and cross sections, Whareama at Waiteko.

n = 0.055

23104: Ngaruroro at Kuripapango.

Map reference:-	U20:969974 (Metric); N123:783533 (Yard).
Catchment area:-	370 km².
Period of record:-	October 1963 - December 1989.
Mean annual flood:-	192 m³/s.
Mean flow:-	17.2 m³/s.

Surveyed reach:-

Cross-sections:-	2 along a 76 m reach.
Manning's n range:-	0.045-0.062
Channel description:-	Bed consists of cobbles and small boulders. Banks are lined with brush and scrub.

Bed Surface Material

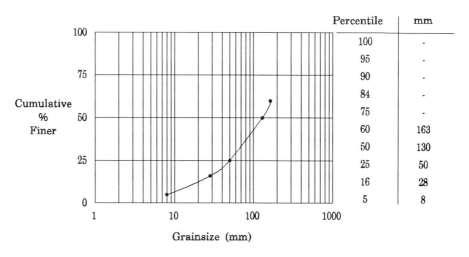

Percentile	mm
100	-
95	-
90	-
84	-
75	-
60	163
50	130
25	50
16	28
5	8

View upstream towards top cross-section.

View downstream towards bottom cross-section.

n = 0.055

Hydraulic Properties of Reach

Discharge	Water Surface Slope	Friction Slope	Area	Expansion	Hydraulic Radius	Mean Velocity	Manning n	Chezy C	Error
(m³/s)			(m²)	(%)	(m)	(m/s)			(%)
9.90*	0.00482	0.00472	14.7	-14	0.47	0.68	0.062	14.3	8
10.8*	0.00491	0.00480	15.2	-15	0.49	0.71	0.060	14.7	8
12.7*	0.00504	0.00491	16.3	-15	0.52	0.78	0.058	15.5	8
15.5*	0.00513	0.00498	17.8	-14	0.56	0.87	0.055	16.6	8
16.5*	0.00517	0.00500	18.3	-14	0.57	0.91	0.054	17.0	8
21.2*	0.00530	0.00510	20.5	-13	0.63	1.04	0.051	18.3	8
27.9*	0.00555	0.00528	23.3	-13	0.71	1.20	0.048	19.7	8
38.7*	0.00611	0.00565	27.1	-15	0.80	1.44	0.045	21.3	9
60.8*	0.00620	0.00572	35.6	-11	1.03	1.72	0.045	22.4	9

* Estimated from rating based on gaugings

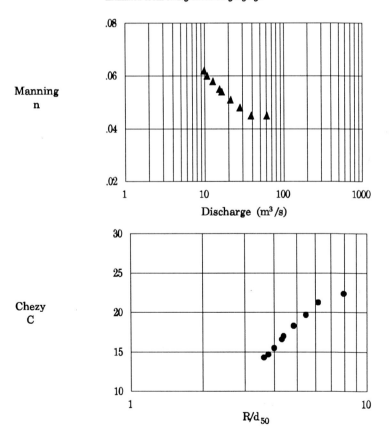

220

$$n = 0.055$$

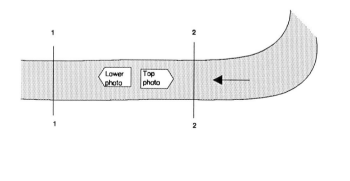

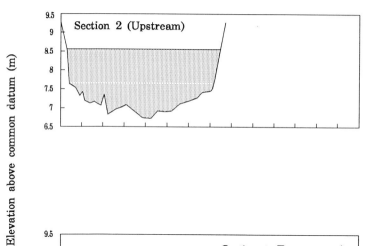

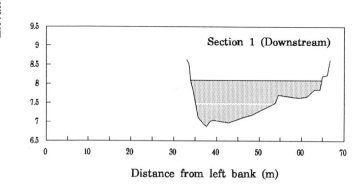

Plan (not to scale) and cross sections, Ngaruroro at Kuripapango.

n = 0.056

15410: Whirinaki at Galatea.

Map reference:-	V17:370960 (Metric); N086:191623 (Yard).
Catchment area:-	534 km².
Period of record:-	December 1952 - December 1989.
Mean annual flood:-	109 m³/s.
Mean flow:-	14.8 m³/s.

Surveyed reach:-

Cross-sections:-	4 along a 338 m reach.
Manning's n range:-	0.037-0.070
Channel description:-	Bed material is mainly cobbles but with some boulders and patches of sand-gravel. Banks, formed of similar material, are lined with overhanging willows and a dense scrub of juvenile willows.

Bed Surface Material

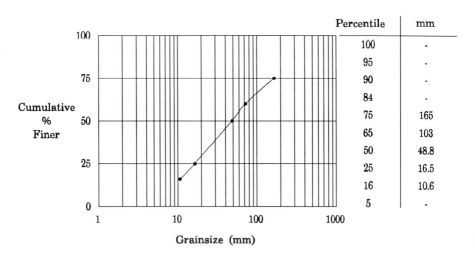

Percentile	mm
100	-
95	-
90	-
84	-
75	165
65	103
50	48.8
25	16.5
16	10.6
5	-

Grainsize (mm)

$n = 0.056$

View upstream from cross-section 2.

View downstream from cross-section 2.

n = 0.056

Hydraulic Properties of Reach

Discharge	Water Surface Slope	Friction Slope	Area	Expansion	Hydraulic Radius	Mean Velocity	Manning n	Chezy C	Error
(m³/s)			(m²)	(%)	(m)	(m/s)			(%)
6.59	0.00469	0.00468	11.7	9	0.55	0.57	0.070	12.6	8
22.9*	0.00441	0.00442	20.3	23	0.79	1.14	0.047	20.4	8
34.3	0.00443	0.00444	24.9	22	0.93	1.39	0.044	22.5	8
35.3	0.00454	0.00454	25.9	18	0.97	1.37	0.046	21.7	8
61.8	0.00461	0.00463	33.1	15	1.18	1.88	0.039	26.4	8
64.0	0.00474	0.00471	32.8	12	1.18	1.97	0.037	27.8	8

* Estimated from rating based on gaugings

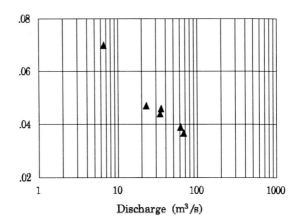

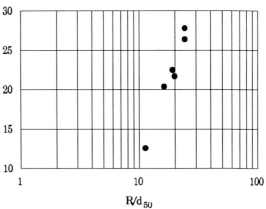

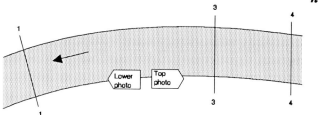

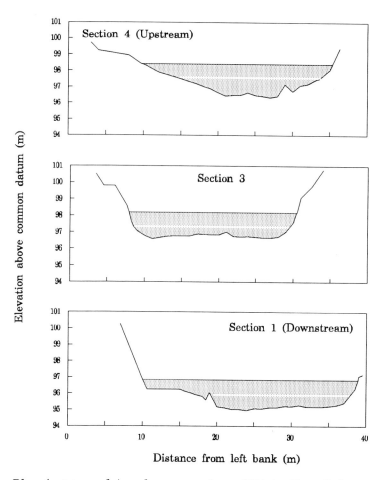

Plan (not to scale) and cross sections, Whirinaki at Galatea.

n = 0.058

1903: Oruru at Saleyards.

Map reference:-	OO4:575801 (Metric); N010:016737 (Yard).
Catchment area:-	79 km².
Period of record:-	December 1988 - Present.
Mean annual flood:-	57 m³/s.
Mean flow:-	3.72 m³/s.

Surveyed reach:-

Cross-sections:-	3 along a 100 m reach.
Manning's n range:-	0.046-0.075
Channel description:-	Bed consists of gravel and small cobbles. Both banks are lined with grazed coarse grass.

Bed Surface Material

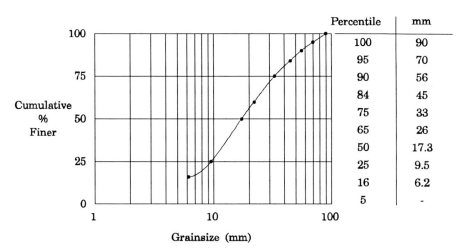

Percentile	mm
100	90
95	70
90	56
84	45
75	33
65	26
50	17.3
25	9.5
16	6.2
5	-

View upstream from middle of reach.

View downstream from middle of reach.

n = 0.058

Hydraulic Properties of Reach

Discharge	Water Surface Slope	Friction Slope	Area	Expansion	Hydraulic Radius	Mean Velocity	Manning n	Chezy C	Error
(m³/s)			(m²)	(%)	(m)	(m/s)			(%)
1.57*	0.00100	0.00102	5.06	38	0.59	0.31	0.075	12.2	9
2.80*	0.00109	0.00111	6.67	20	0.69	0.42	0.063	15.0	9
5.37	0.00106	0.00107	9.60	5	0.91	0.56	0.055	18.0	9
9.19*	0.00105	0.00103	13.8	-4	1.16	0.67	0.053	19.3	9
12.1	0.00084	0.00080	16.5	-5	1.29	0.73	0.046	22.7	10
14.2	0.00135	0.00128	18.8	-10	1.38	0.76	0.059	18.0	9

* Estimated from rating based on gaugings

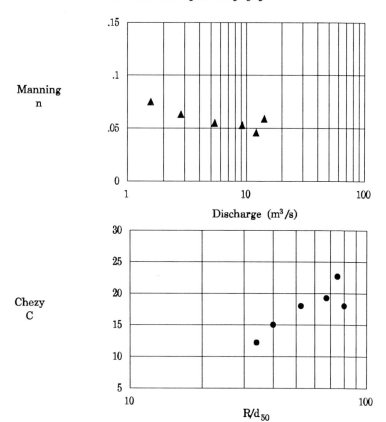

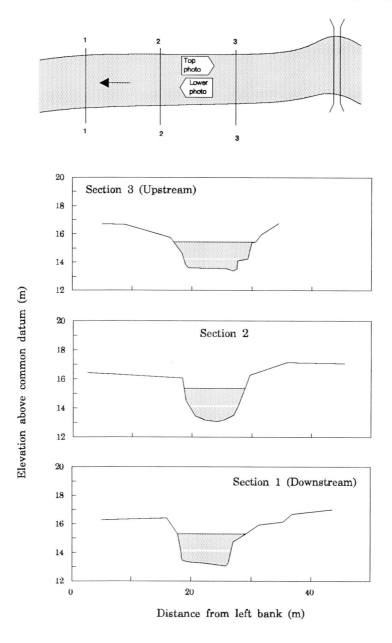

$n = 0.058$

Plan (not to scale) and cross sections, Oruru at Saleyards.

n = 0.059

1316: Awanui at School Cut.

Map reference:-	O04:352761 (Metric); N010:774686 (Yard).
Catchment area:-	222 km².
Period of record:-	January 1958 - Present.
Mean annual flood:-	148 m³/s.
Mean flow:-	6.21 m³/s.

Surveyed reach:-

Cross-sections:-	3 along a 187 m reach.
Manning's n range:-	0.058-0.081
Channel description:-	Channel bed is clay and silt. Both banks are lined with a dense growth of small trees.

View upstream from middle cross-section.

View downstream from middle cross-section.

n = 0.059

Hydraulic Properties of Reach

Discharge	Water Surface Slope	Friction Slope	Area	Expansion	Hydraulic Radius	Mean Velocity	Manning n	Chezy C	Error
(m³/s)			(m²)	(%)	(m)	(m/s)			(%)
8.8*	0.00183	0.00179	13.3	-12	1.08	0.67	0.062	16.1	8
10.8*	0.00153	0.00150	15.3	-6	1.15	0.70	0.058	17.5	8
13.6	0.00157	0.00153	19.3	-12	1.33	0.71	0.065	16.1	8
22.7	0.00109	0.00105	30.5	-13	1.69	0.75	0.061	17.8	8
25.3	0.00112	0.00108	35.6	-13	1.81	0.71	0.069	16.1	8
47.5*	0.00088	0.00083	63.8	-14	2.47	0.75	0.071	16.5	8
49.5*	0.00092	0.00087	65.4	-14	2.51	0.76	0.072	16.2	8
56.0	0.00091	0.00086	70.7	-15	2.62	0.80	0.070	16.8	9
115	0.00137	0.00122	106	-20	3.29	1.09	0.071	17.2	9
143	0.00151	0.00133	122	-22	3.54	1.18	0.072	17.2	9
172	0.00198	0.00175	139	-24	3.74	1.28	0.081	15.4	9

* Estimated from rating based on gaugings

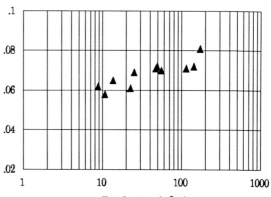

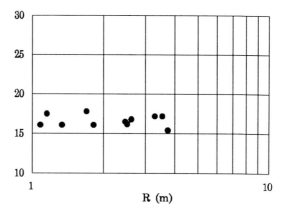

$n = 0.059$

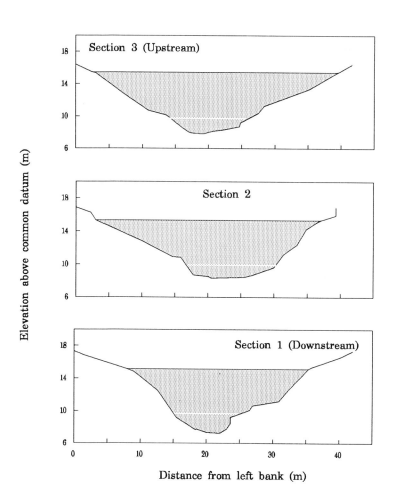

Plan (not to scale) and cross sections, Awanui at School Cut.

233

n = 0.060 (est.)

45311: Kaipara at Waimauku.

Map reference:-	Q10:436919 (Metric); N037:015706 (Yard).
Catchment area:-	155 km².
Period of record:-	October 1978 - Present.
Mean annual flood:-	76 m³/s.
Mean flow:-	2.72 m³/s.

Surveyed reach:-

Cross-sections:-	4 along a 269 m reach.
Manning's n range:-	0.051-0.061
Channel description:-	Bed is clay. Banks are silt with a grass and weed cover.

View from middle of reach looking upstream.

View from cross-section 2 looking downstream.

n = 0.060 (est.)

Hydraulic Properties of Reach

Discharge	Water Surface Slope	Friction Slope	Area	Expansion	Hydraulic Radius	Mean Velocity	Manning n	Chezy C	Error
(m³/s)			(m²)	(%)	(m)	(m/s)			(%)
13.6*	0.00045	0.00043	26.3	10	1.57	0.55	0.056	19.4	9
26.4*	0.00050	0.00048	45.4	8	1.77	0.60	0.057	19.3	9
34.8*	0.00049	0.00048	58.7	9	2.04	0.61	0.061	18.5	9
35.4*	0.00048	0.00047	59.1	9	2.05	0.62	0.060	18.9	9
36.2*	0.00049	0.00048	60.0	10	2.07	0.62	0.061	18.6	9
72.0*	0.00059	0.00057	83.9	10	2.43	0.88	0.051	22.6	9

* Estimated from rating based on gaugings

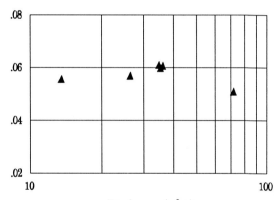

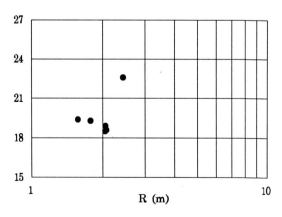

$n = 0.060$ *(est.)*

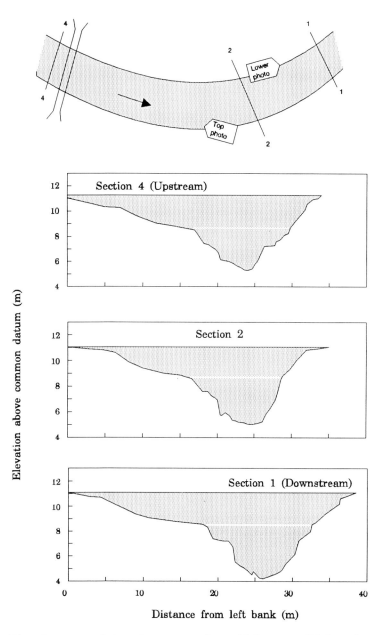

Plan (not to scale) and cross sections, Kaipara at Waimauku.

n = 0.061

47804: Waipapa at Forest Ranger.

Map reference:-	P05:730583 (Metric); N010:193503 (Yard).
Catchment area:-	122 km².
Period of record:-	September 1975 - Present.
Mean annual flood:-	250 m³/s.
Mean flow:-	4.59 m³/s.

Surveyed reach:-

Cross-sections:-	3 along a 182 m reach.
Manning's n range:-	0.033-0.066
Channel description:-	Bed material consists of small and large cobbles. Both banks are covered with small native trees, brush, and grasses.

Bed Surface Material

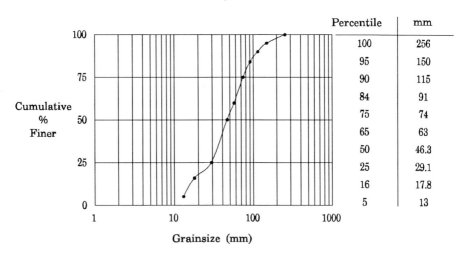

Percentile	mm
100	256
95	150
90	115
84	91
75	74
65	63
50	46.3
25	29.1
16	17.8
5	13

View upstream from middle cross-section.

View downstream from middle cross-section.

_n = 0.061_

Hydraulic Properties of Reach

Discharge (m³/s)	Water Surface Slope	Friction Slope	Area (m²)	Expansion (%)	Hydraulic Radius (m)	Mean Velocity (m/s)	Manning n	Chezy C	Error (%)
4.55*	0.00269	0.00268	10.2	-9	0.39	0.45	0.063	13.6	8
4.73*	0.00282	0.00280	10.8	-14	0.41	0.44	0.066	13.0	8
10.6*	0.00255	0.00255	15.6	-4	0.59	0.68	0.052	17.5	8
43.2*	0.00179	0.00183	31.6	9	1.10	1.37	0.033	30.6	9
59.3	0.00185	0.00183	37.7	5	1.28	1.57	0.033	32.0	9

* Estimated from rating based on gaugings

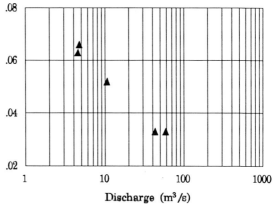

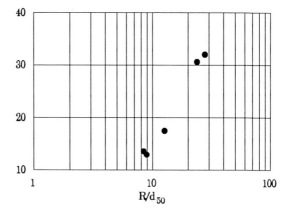

$$n = 0.061$$

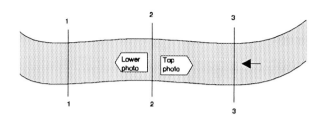

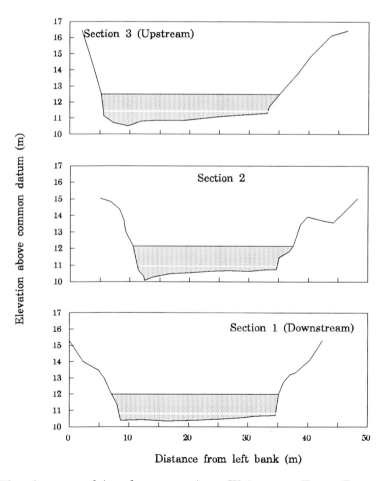

Plan (not to scale) and cross sections, Waipapa at Forest Ranger.

n = 0.061

75259: Fraser at Old Man Range.

Map reference:-	G42:122472 (Metric); S25:031485 (Yard).
Catchment area:-	122 km².
Period of record:-	May 1969 - Present.
Mean annual flood:-	35 m³/s.
Mean flow:-	2.17 m³/s.

Surveyed reach:-

Cross-sections:-	3 along a 102 m reach.
Manning's n range:-	0.052-0.061
Channel description:-	Bed material ranges from fine gravel to boulders (up to 1m in diameter). Left bank is covered in bush and tussock. Right bank is mainly small boulders and gravel, the remnants of gold mining.

Bed Surface Material

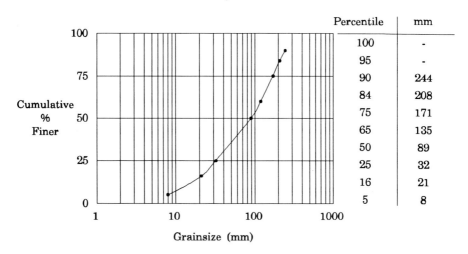

Percentile	mm
100	-
95	-
90	244
84	208
75	171
65	135
50	89
25	32
16	21
5	8

$n = 0.061$

n = 0.061

Hydraulic Properties of Reach

Discharge (m³/s)	Water Surface Slope	Friction Slope	Area (m²)	Expansion (%)	Hydraulic Radius (m)	Mean Velocity (m/s)	Manning n	Chezy C	Error (%)
2.56	0.00819	0.00807	3.67	-9	0.31	0.71	0.061	13.5	8
4.65*	0.00829	0.00763	4.76	29	0.39	1.00	0.052	16.6	8
5.23	0.00835	0.00830	5.31	11	0.42	1.00	0.054	16.0	8

* Estimated from rating based on gaugings

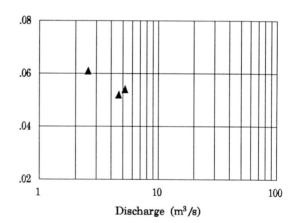

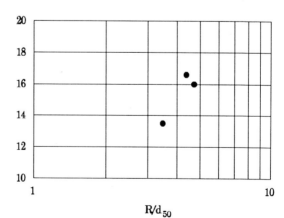

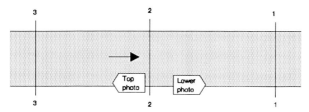

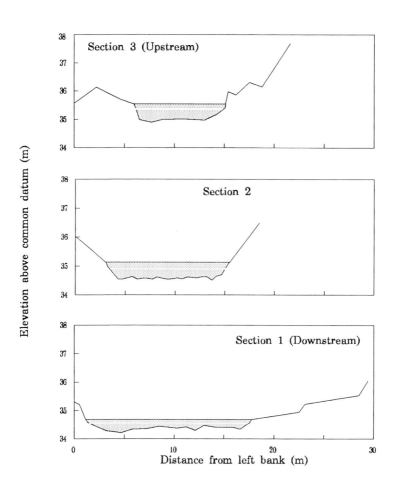

Plan (not to scale) and cross sections, Fraser at Old Man Range.

n = 0.063

1043419: Pokaiwhenau at Puketurua.

Map reference:-	T15:490462 (Metric); N075:213145 (Yard).
Catchment area:-	448 km².
Period of record:-	September 1963 - January 1990.
Mean annual flood:-	23.7 m³/s.
Mean flow:-	5.06 m³/s.

Surveyed reach:-

Cross-sections:-	3 along a 100 m reach.
Manning's n range:-	0.028-0.10
Channel description:-	Bed is mostly bedrock with some cobbles and boulders. Bank cover is mainly pasture grasses and some blackberry bushes.

View downstream from top of reach.

View upstream from bottom cross-section.

n = 0.063

Hydraulic Properties of Reach

Discharge	Water Surface Slope	Friction Slope	Area	Expansion	Hydraulic Radius	Mean Velocity	Manning n	Chezy C	Error
(m³/s)			(m²)	(%)	(m)	(m/s)			(%)
2.78	0.00641	0.00649	6.45	87	0.47	0.49	0.101	8.5	8
4.11	0.00625	0.00641	7.11	85	0.50	0.68	0.080	11.0	8
8.35	0.00446	0.00495	9.01	108	0.61	1.03	0.046	19.4	8
26.3	0.00127	0.00230	18.4	105	1.05	1.57	0.030	33.0	14
30.0	0.00108	0.00144	21.8	42	1.11	1.40	0.028	36.4	13

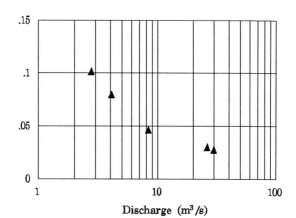

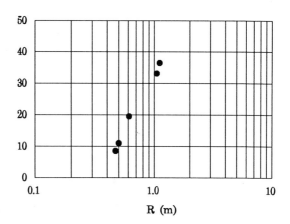

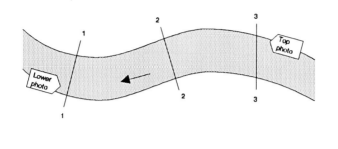

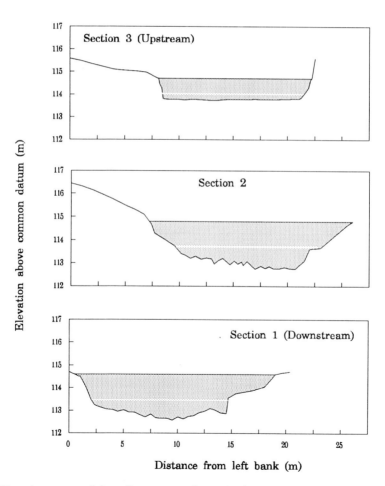

Plan (not to scale) and cross sections, Pokaiwhenau at Puketurua.

249

n = 0.065

29250: Ruakokapatuna at Iraia.

Map reference:-	S28:085778 (Metric); N165:884105 (Yard).
Catchment area:-	15.5 km².
Period of record:-	May 1969 - December 1988.
Mean annual flood:-	34 m³/s.
Mean flow:-	0.7 m³/s.

Surveyed reach:-

Cross-sections:-	3 along a 70 m reach.
Manning's n range:-	0.038-0.15
Channel description:-	Bed consists of gravel and cobbles. Both banks are grass covered, with the right bank having some overhanging trees.

Bed Surface Material

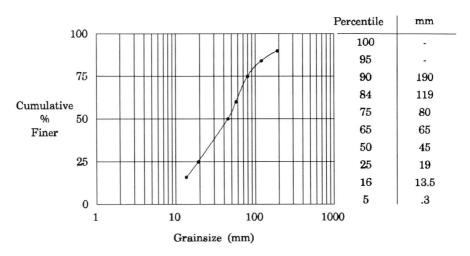

Percentile	mm
100	-
95	-
90	190
84	119
75	80
65	65
50	45
25	19
16	13.5
5	.3

$$n = 0.065$$

View downstream from top cross-section.

View upstream from middle of reach.

251

n = 0.065

Hydraulic Properties of Reach

Discharge	Water Surface Slope	Friction Slope	Area	Expansion	Hydraulic Radius	Mean Velocity	Manning n	Chezy C	Error
(m³/s)			(m²)	(%)	(m)	(m/s)			(%)
0.08*	0.00559	0.00559	0.62	-14	0.13	0.12	0.148	4.80	8
0.20*	0.00569	0.00568	0.98	0	0.17	0.22	0.100	7.40	8
0.22*	0.00572	0.00570	0.84	-20	0.14	0.26	0.075	9.50	8
0.29*	0.00522	0.00523	1.14	20	0.16	0.26	0.084	8.80	8
0.41*	0.00558	0.00557	1.35	7	0.19	0.31	0.073	10.2	8
0.89*	0.00529	0.00533	1.87	24	0.24	0.48	0.057	13.7	8
3.89*	0.00539	0.00545	3.97	8	0.42	0.98	0.042	20.5	8
6.63*	0.00608	0.00612	5.44	6	0.53	1.22	0.042	21.5	8
10.0*	0.00688	0.00641	6.92	-14	0.63	1.45	0.040	23.2	9
10.9*	0.00688	0.00634	7.13	-14	0.64	1.54	0.038	24.4	9
15.2*	0.00783	0.00706	9.58	-17	0.78	1.60	0.043	22.1	9

* Estimated from rating based on gaugings

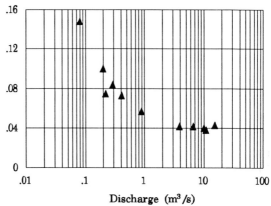

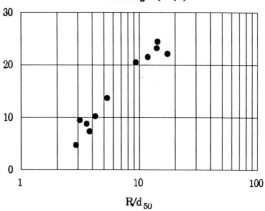

252

$$n = 0.065$$

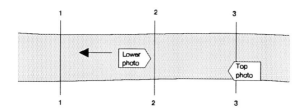

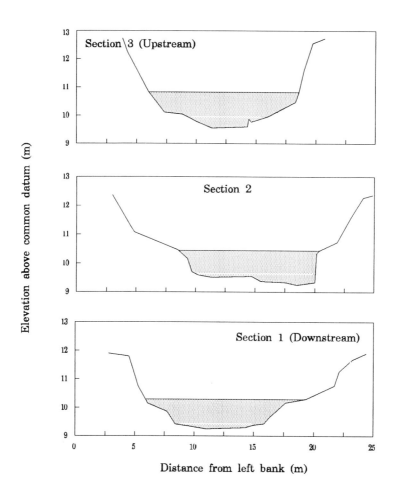

Plan (not to scale) and cross sections, Ruakokapatuna at Iraia.

253

n = 0.066

80201: Rowallanburn at Old Mill.

Map reference:-	C46:879382 (Metric); S167:650316 (Yard).
Catchment area:-	71.6 km².
Period of record:-	December 1969 - 1980.
Mean annual flood:-	27.1 m³/s.
Mean flow:-	1.29 m³/s.

Surveyed reach:-

Cross-sections:-	3 along a 40 m reach.
Manning's n range:-	0.029-0.098
Channel description:-	Bed comprises cobbles and boulders. Banks are lined with overhanging ferns and native bush.

Bed Surface Material

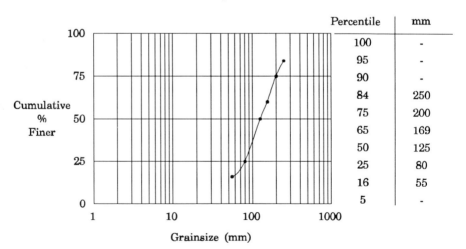

Percentile	mm
100	-
95	-
90	-
84	250
75	200
65	169
50	125
25	80
16	55
5	-

View upstream towards top of reach.

View downstream at bottom cross-section.

n = 0.066

Hydraulic Properties of Reach

Discharge	Water Surface Slope	Friction Slope	Area	Expansion	Hydraulic Radius	Mean Velocity	Manning n	Chezy C	Error
(m³/s)			(m²)	(%)	(m)	(m/s)			(%)
0.50	0.00027	0.00026	5.28	-24	0.62	0.10	0.098	9.2	30
3.57	0.00045	0.00031	7.90	-21	0.86	0.47	0.030	32.6	27
8.77*	0.00190	0.00189	10.6	17	1.07	0.84	0.050	19.9	11
11.1*	0.00160	0.00112	11.5	-15	1.13	0.98	0.035	28.9	17
14.7*	0.00152	0.00091	13.0	-14	1.25	1.14	0.029	35.3	25

* Estimated from rating based on gaugings

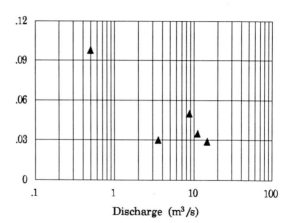

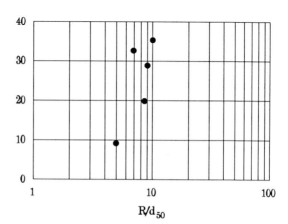

256

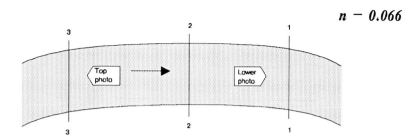

$n - 0.066$

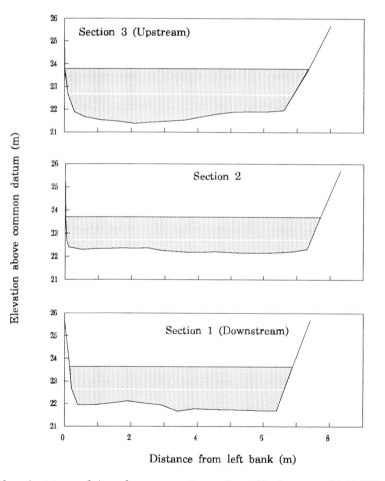

Plan (not to scale) and cross sections, Rowallanburn at Old Mill.

n = 0.07

52916: Cobb at Trilobite.

Map reference:-	M27:773088 (Metric); S013:020482 (Yard).
Catchment area:-	46.8 km².
Period of record:-	May 1969 - January 1990.
Mean annual flood:-	97 m³/s.
Mean flow:-	3.69 m³/s.

Surveyed reach:-

Cross-sections:-	4 along a 123 m reach.
Manning's n range:-	0.038-0.13
Channel description:-	Bed consists mainly of gravel with some boulders. Banks are lined with tussock and occasional low scrub. Some undercut grassy banks.

Bed Surface Material

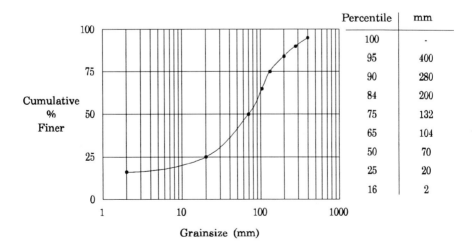

Percentile	mm
100	-
95	400
90	280
84	200
75	132
65	104
50	70
25	20
16	2

View downstream from top cross-section.

View upstream from bottom cross-section.

n = 0.07

Hydraulic Properties of Reach

Discharge	Water Surface Slope	Friction Slope	Area	Expansion	Hydraulic Radius	Mean Velocity	Manning n	Chezy C	Error
(m³/s)			(m²)	(%)	(m)	(m/s)			(%)
0.78*	0.00153	0.00153	5.12	54	0.44	0.15	0.130	6.5	8
1.39*	0.00154	0.00155	6.12	45	0.40	0.23	0.092	9.3	8
5.96*	0.00207	0.00207	10.4	0	0.63	0.57	0.058	15.9	8
32.1*	0.00302	0.00285	21.6	-8	1.12	1.49	0.038	26.5	9
40.4*	0.00320	0.00297	26.5	-7	1.21	1.53	0.040	25.7	9

* Estimated from rating based on gaugings

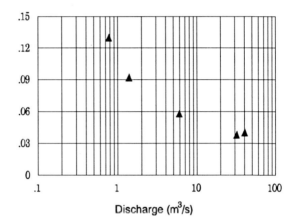

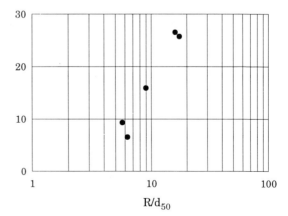

260

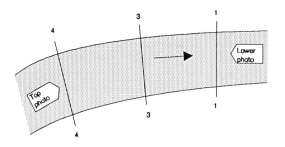

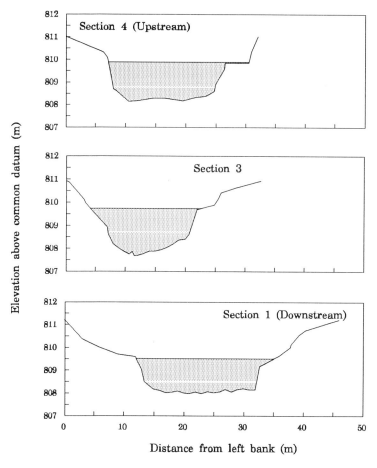

Plan (not to scale) and cross sections, Cobb at Trilobite.

261

n = 0.085

34305: Patea at McColls Bridge.

Map reference:-	Q21:429757 (Metric); N130:105248 (Yard).
Catchment area:-	900 km².
Period of record:-	November 1986 - 1989.
Mean annual flood:-	364 m³/s.
Mean flow:-	36.9 m³/s.

Surveyed reach:-

Cross-sections:-	5 along a 400 m reach.
Manning's n range:-	0.045-0.18
Channel description:-	Bed is mainly sand with some gravel and scattered submerged logs. Banks are covered with clean grass.

Bed Surface Material

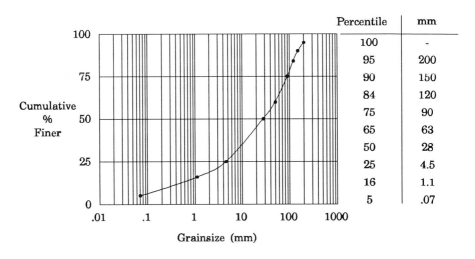

Percentile	mm
100	-
95	200
90	150
84	120
75	90
65	63
50	28
25	4.5
16	1.1
5	.07

View upstream from cross-section 3.

View downstream from cross-section 3.

n = 0.085

Hydraulic Properties of Reach

Discharge	Water Surface Slope	Friction Slope	Area	Expansion	Hydraulic Radius	Mean Velocity	Manning n	Chezy C	Error
(m³/s)			(m²)	(%)	(m)	(m/s)			(%)
2.80*	0.00027	0.00027	27.3	107	1.30	0.12	0.18	5.74	8
46.0*	0.00143	0.00141	57.0	22	2.10	0.83	0.078	14.6	8
61.0*	0.00146	0.00144	65.6	19	2.32	0.95	0.073	15.9	8
62.0*	0.00151	0.00149	66.1	18	2.32	0.96	0.074	15.7	8
130*	0.00110	0.00110	97.0	22	2.97	1.35	0.052	23.3	9
218*	0.00102	0.00103	132	22	3.51	1.66	0.045	27.6	10

* Estimated from rating based on gaugings

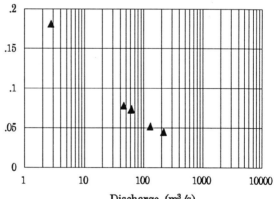

Manning n

Discharge (m³/s)

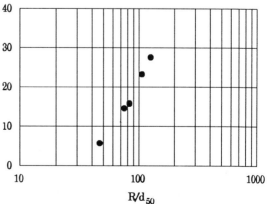

Chezy C

R/d_{50}

264

$n = 0.085$

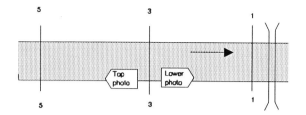

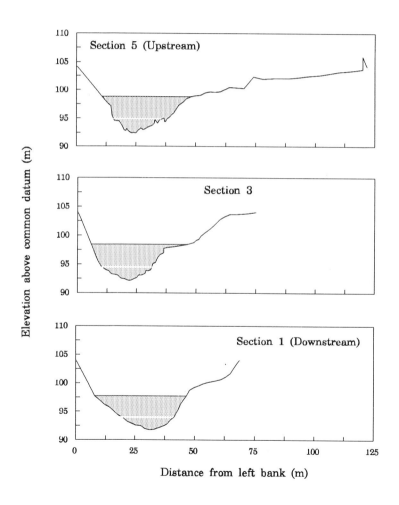

Plan (not to scale) and cross sections, Patea at McColls Bridge.

n = 0.085

664042: Northbrook at Oxidation Pond.

Map reference:-	M35:782660 (Metric); S076:986827 (Yard).
Catchment area:-	Artesian source.
Period of record:-	December 1988 - May 1989.
Mean annual flood:-	1 m³/s.
Mean flow:-	0.121 m³/s.

Surveyed reach:-

Cross-sections:-	2 along a 24 m reach.
Manning's n range:-	0.072-0.091
Channel description:-	Bed is composed mainly of gravel and cobbles with some mud. Banks are lined with long uncut grass.

Bed Surface Material

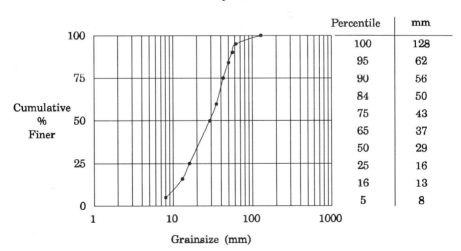

Percentile	mm
100	128
95	62
90	56
84	50
75	43
65	37
50	29
25	16
16	13
5	8

View downstream at bottom cross-section.

View upstream at top cross-section.

n = 0.085

Hydraulic Properties of Reach

Discharge	Water Surface Slope	Friction Slope	Area	Expansion	Hydraulic Radius	Mean Velocity	Manning n	Chezy C	Error
(m³/s)			(m²)	(%)	(m)	(m/s)			(%)
0.10*	0.00630	0.00612	0.45	-50	0.16	0.26	0.091	8.0	8
0.12*	0.00630	0.00601	0.50	-57	0.16	0.29	0.087	8.5	8
0.13	0.00634	0.00601	0.53	-57	0.17	0.29	0.088	8.4	8
0.16	0.00638	0.00612	0.57	-50	0.19	0.31	0.086	8.8	8
0.21	0.00643	0.00775	0.67	-44	0.20	0.34	0.083	9.2	8
0.26	0.00630	0.00776	0.45	-46	0.22	0.26	0.091	8.0	8
0.33*	0.00655	0.00627	0.87	-36	0.25	0.40	0.078	10.1	8
0.46*	0.00677	0.00637	1.02	-33	0.28	0.47	0.074	11.0	8
0.59*	0.00681	0.00644	1.19	-29	0.31	0.51	0.072	11.4	8
0.69*	0.00681	0.00641	1.33	-27	0.33	0.54	0.072	11.5	8

* Estimated from rating based on gaugings

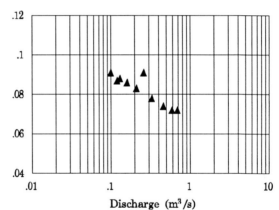

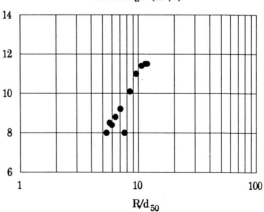

$$n = 0.085$$

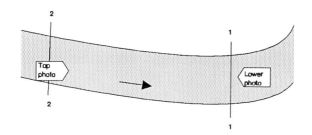

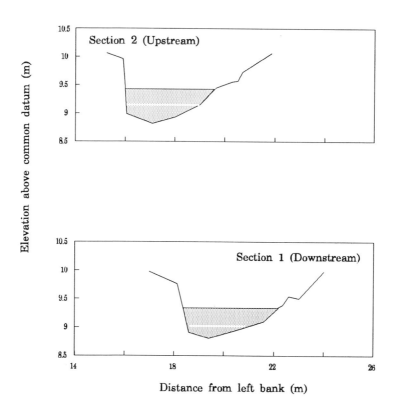

Plan (not to scale) and cross sections, Northbrook at Oxidation Pond.

269

$n = 0.088$

58902: Pelorus at Bryants.

Map reference:-	O27:573891 (Metric); S021:891251 (Yard).
Catchment area:-	375 km².
Period of record:-	October 1977 - January 1990.
Mean annual flood:-	957 m³/s.
Mean flow:-	20.8 m³/s.

Surveyed reach:-

Cross-sections:-	5 along a 472 m reach.
Manning's n range:-	0.021-0.17
Channel description:-	Bed consists predominantly of large cobbles with some small cobbles and gravel. Bedrock crops out on both banks, which are lined with native bush towards the upstream end of the reach. Left bank is grassed along lower section of reach.

Bed Surface Material

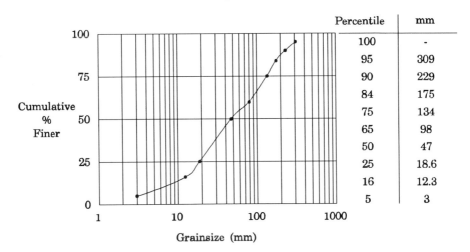

Percentile	mm
100	-
95	309
90	229
84	175
75	134
65	98
50	47
25	18.6
16	12.3
5	3

View upstream towards top cross-section.

View downstream towards bottom cross-section.

n = 0.088

Hydraulic Properties of Reach

Discharge	Water Surface Slope	Friction Slope	Area	Expansion	Hydraulic Radius	Mean Velocity	Manning n	Chezy C	Error
(m³/s)			(m²)	(%)	(m)	(m/s)			(%)
6.13*	0.00366	0.00364	26.9	-51	0.76	0.27	0.17	5.6	8
11.8*	0.00356	0.00352	31.7	-50	0.83	0.44	0.11	8.9	8
13.1*	0.00381	0.00375	33.2	-56	0.82	0.47	0.11	8.9	8
27.3*	0.00415	0.00390	42.7	-66	0.97	0.78	0.074	13.1	8
79.7*	0.00337	0.00273	66.3	-60	1.41	1.38	0.048	21.9	8
164*	0.00326	0.00183	92.3	-62	1.84	2.04	0.032	34.2	11
290*	0.00333	0.00102	124	-65	2.25	2.65	0.021	54.0	22

* Estimated from rating based on gaugings

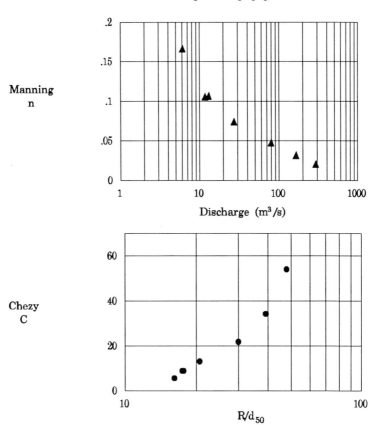

272

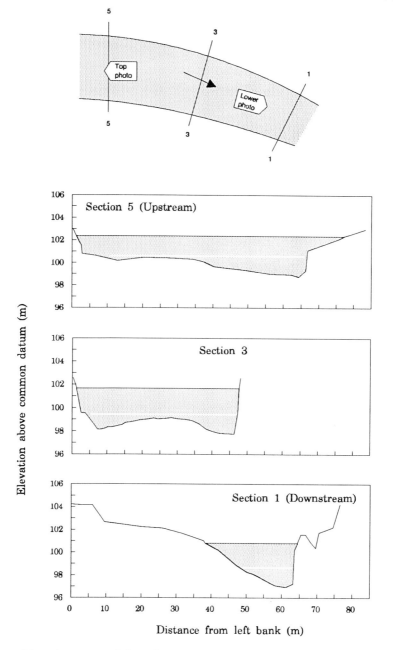

$n = 0.088$

Section 5 (Upstream)

Section 3

Section 1 (Downstream)

Elevation above common datum (m)

Distance from left bank (m)

Plan (not to scale) and cross sections, Pelorus at Bryants.

273

n = 0.088

93901: Ngakawau at Lineslip.

Map reference:-	L28:177548 (Metric); S024:358903 (Yard).
Catchment area:-	186 km².
Period of record:-	June 1974 - Present.
Mean annual flood:-	676 m³/s.
Mean flow:-	25.6 m³/s.

Surveyed reach:-

Cross-sections:-	5 along a 462 m reach.
Manning's n range:-	0.057-0.25
Channel description:-	Bed material comprises 50% boulders and 50% sand and cobbles. Banks are formed of sand and boulders and are bush covered.

Bed Surface Material

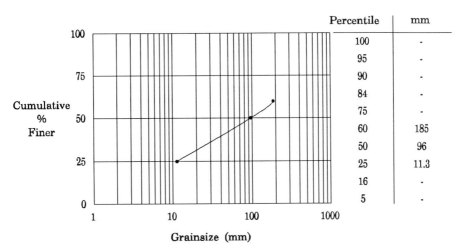

Percentile	mm
100	-
95	-
90	-
84	-
75	-
60	185
50	96
25	11.3
16	-
5	-

View downstream from middle of reach.

View upstream from bottom of reach.

n = 0.088

Hydraulic Properties of Reach

Discharge	Water Surface Slope	Friction Slope	Area	Expansion	Hydraulic Radius	Mean Velocity	Manning n	Chezy C	Error
(m³/s)			(m²)	(%)	(m)	(m/s)			(%)
1.64	0.00330	0.00329	18.6	-63	0.72	0.12	0.25	3.5	8
2.50	0.00334	0.00333	20.0	-60	0.75	0.16	0.19	4.5	8
16.7*	0.00372	0.00369	35.5	-35	1.01	0.51	0.096	9.8	8
19.5*	0.00363	0.00358	35.5	-32	1.01	0.60	0.079	12.0	8
39.5*	0.00399	0.00392	45.1	-26	1.16	0.91	0.064	15.4	8
40.3*	0.00416	0.00411	49.6	-25	1.23	0.83	0.077	12.9	8
94.1*	0.00460	0.00446	66.1	-18	1.50	1.44	0.057	18.3	8

* Estimated from rating based on gaugings

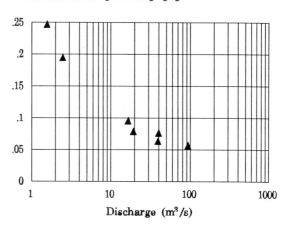

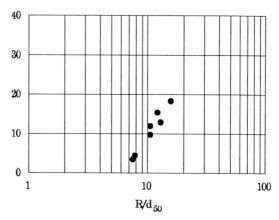

$n = 0.088$

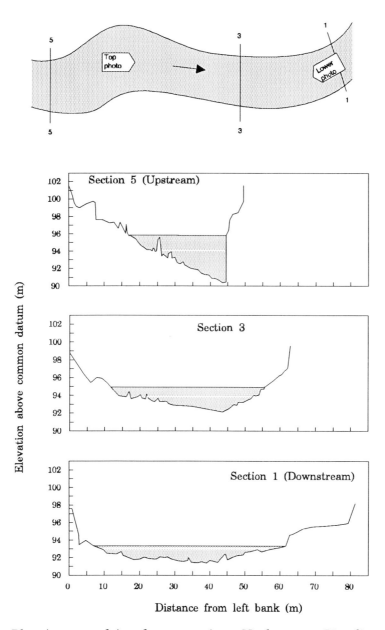

Plan (not to scale) and cross sections, Ngakawau at Lineslip.

n = 0.096

37503: Kapoaiaia at Lighthouse.

Map reference:-	P20:755143 (Metric); N118:356649 (Yard).
Catchment area:-	20 km².
Period of record:-	February 1986 - 1989.
Mean annual flood:-	23 m³/s.
Mean flow:-	1.06 m³/s.

Surveyed reach:-

Cross-sections:-	5 along a 99 m reach.
Manning's n range:-	0.061-0.11
Channel description:-	Bed is mainly gravel and cobbles, but with small boulders scattered along the reach. Right bank is clean grass, while the left bank has longer grass and some overhanging scrub.

Bed Surface Material

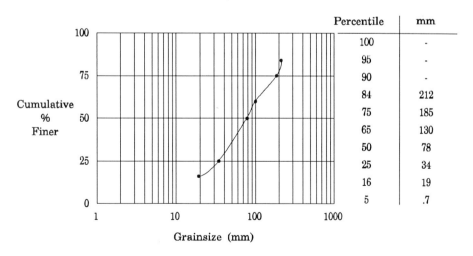

Percentile	mm
100	-
95	-
90	-
84	212
75	185
65	130
50	78
25	34
16	19
5	.7

$$n = 0.096$$

View upstream to top cross-section.

View downstream to bottom cross-section.

n = 0.096

Hydraulic Properties of Reach

Discharge	Water Surface Slope	Friction Slope	Area	Expansion	Hydraulic Radius	Mean Velocity	Manning n	Chezy C	Error
(m³/s)			(m²)	(%)	(m)	(m/s)			(%)
0.49*	0.00947	0.00944	1.67	-25	0.22	0.30	0.11	7.2	8
0.66*	0.00935	0.00933	2.04	-13	0.26	0.33	0.11	7.1	8
1.25*	0.00962	0.00962	2.74	-10	0.33	0.46	0.096	8.6	8
1.62*	0.00972	0.00964	2.85	-13	0.34	0.58	0.079	10.5	8
3.10*	0.00987	0.00980	4.26	-7	0.48	0.73	0.081	10.9	8
5.20*	0.01090	0.01070	4.89	-19	0.54	1.07	0.061	14.7	8

* Estimated from rating based on gaugings

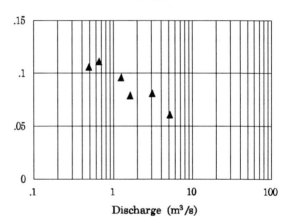

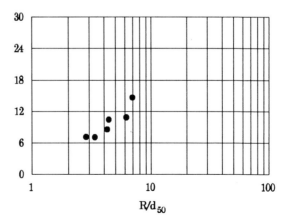

$n - 0.096$

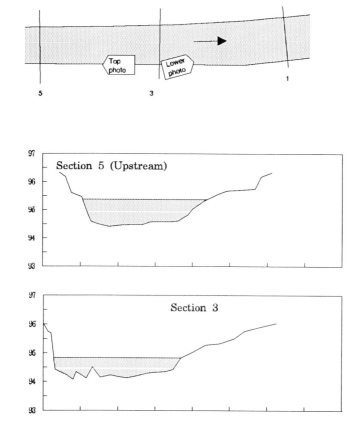

Plan (not to scale) and cross sections, Kapoaiaia at Lighthouse.

n = 0.12

30516: Mill Creek at Papanui.

Map reference:-	R27:589017 (Metric); N160:334350 (Yard).
Catchment area:-	9.35 km².
Period of record:-	April 1969 - December 1988.
Mean annual flood:-	7.9 m³/s.
Mean flow:-	0.14 m³/s.

Surveyed reach:-

Cross-sections:-	3 along a 36 m reach.
Manning's n range:-	0.056-0.25
Channel description:-	Bed consists of mud and silt with a few areas of fine gravel. Banks are grass covered and extend on to paddocks on both sides.

Bed Surface Material

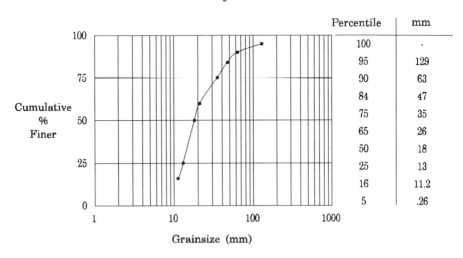

Percentile	mm
100	-
95	129
90	63
84	47
75	35
65	26
50	18
25	13
16	11.2
5	.26

n = 0.12

View upstream from bottom cross-section.

n = 0.12

Hydraulic Properties of Reach

Discharge	Water Surface Slope	Friction Slope	Area	Expansion	Hydraulic Radius	Mean Velocity	Manning n	Chezy C	Error
(m³/s)			(m²)	(%)	(m)	(m/s)			(%)
0.01*	0.00159	0.00156	0.53	-89	0.18	0.06	0.25	2.8	13
0.02*	0.00030	0.00025	0.83	-93	0.41	0.06	0.23	3.7	31
0.05*	0.00174	0.00159	0.62	-80	0.20	0.13	0.12	6.2	13
0.26*	0.00173	0.00151	1.36	-71	0.36	0.25	0.074	10.9	9
0.29*	0.00373	0.00345	1.56	-74	0.39	0.25	0.119	6.8	8
0.47*	0.00294	0.00255	1.60	-67	0.40	0.36	0.072	11.6	9
0.69*	0.00227	0.00200	1.98	-52	0.46	0.38	0.067	12.9	9
1.08*	0.00333	0.00291	2.30	-48	0.49	0.50	0.064	13.6	9
2.06*	0.00410	0.00365	3.23	-35	0.56	0.65	0.061	17.8	8
2.14*	0.00398	0.00355	3.32	-33	0.56	0.66	0.060	15.1	9
2.34*	0.00415	0.00372	3.52	-31	0.58	0.68	0.061	14.9	8
8.52§	0.00489	0.00475	8.11	-5	0.80	1.05	0.056	17.1	9

* Estimated from rating based on gaugings   § Estimated from theoretical rating for structure

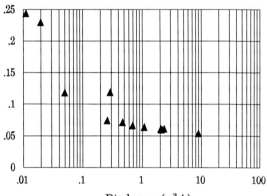

Manning n

Discharge (m³/s)

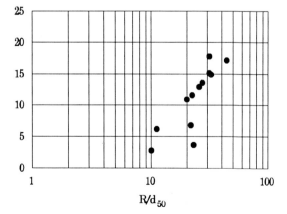

Chezy C

R/d₅₀

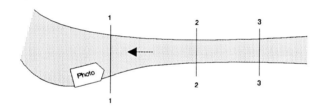

$n = 0.12$

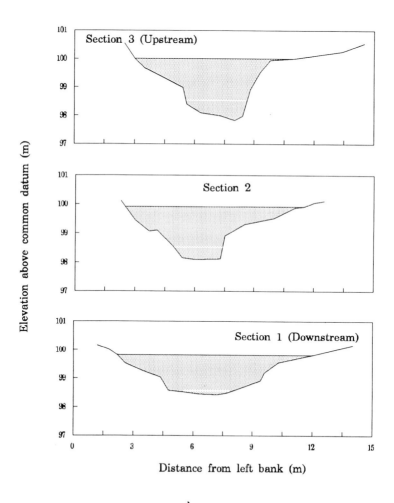

Plan (not to scale) and cross sections, Mill Creek at Papanui.

n = 0.12

4901: Ngunguru at Dugmores Rock.

Map reference:-	Q06:405214 (Metric); N020:942120 (Yard).
Catchment area:-	12.5 km².
Period of record:-	August 1969 - Present.
Mean annual flood:-	61 m³/s.
Mean flow:-	0.41 m³/s.

Surveyed reach:-

Cross-sections:-	3 along a 80 m reach.
Manning's n range:-	0.051-0.16
Channel description:-	Bed consists of gravel and cobbles. Both banks are lined with grazed grass and scattered brush.

Bed Surface Material

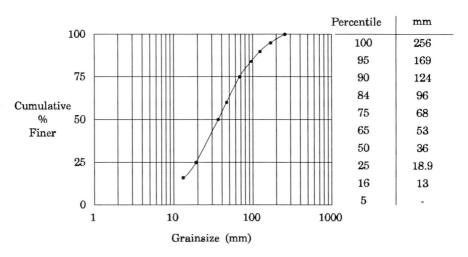

Percentile	mm
100	256
95	169
90	124
84	96
75	68
65	53
50	36
25	18.9
16	13
5	-

View upstream from middle cross-section.

View downstream from middle cross-section.

n = 0.12

Hydraulic Properties of Reach

Discharge	Water Surface Slope	Friction Slope	Area	Expansion	Hydraulic Radius	Mean Velocity	Manning n	Chezy C	Error
(m³/s)			(m²)	(%)	(m)	(m/s)			(%)
0.38*	0.00372	0.00361	3.05	-84	0.42	0.21	0.16	5.4	8
0.61	0.00370	0.00334	3.21	-86	0.43	0.37	0.088	9.3	8
2.19*	0.00475	0.00386	4.71	-76	0.58	0.65	0.066	13.5	8
5.03§	0.00639	0.00489	6.28	-67	0.77	1.08	0.062	15.5	9
7.72§	0.00699	0.00510	8.19	-64	0.81	1.13	0.055	17.2	9
12.2§	0.00651	0.00513	11.3	-53	0.92	1.18	0.055	17.6	9
12.3§	0.00652	0.00493	11.2	-54	0.91	1.22	0.053	18.3	9
17.9§	0.00745	0.00548	13.7	-51	1.00	1.43	0.051	19.4	9
20.2§	0.00758	0.00575	15.1	-49	1.06	1.44	0.054	18.6	9
25.1§	0.00728	0.00581	18.4	-42	1.17	1.43	0.058	17.5	9
29.3§	0.00636	0.00495	20.0	-37	1.23	1.52	0.053	19.4	9

* Estimated from rating based on gaugings § Estimated from theoretical rating for structure

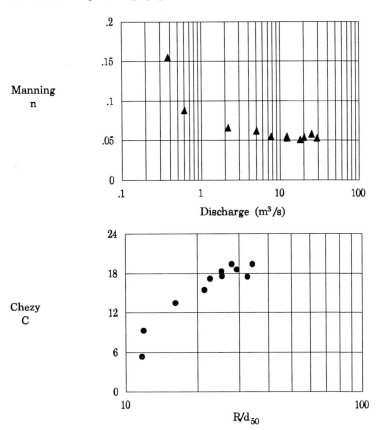

288

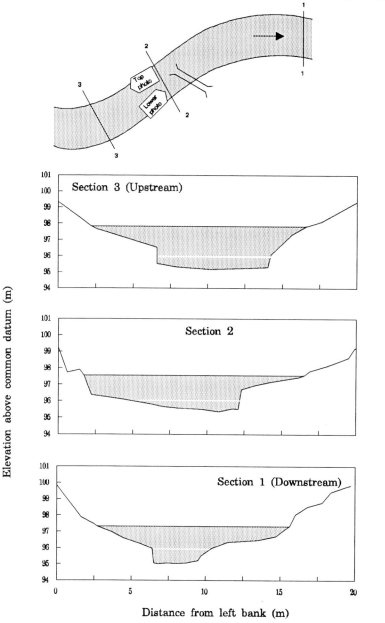

$n = 0.12$

Section 3 (Upstream)

Section 2

Section 1 (Downstream)

Elevation above common datum (m)

Distance from left bank (m)

Plan (not to scale) and cross sections, Ngunguru at Dugmores Rock.

n = 0.12

90605: Butchers Creek at Lake Kaniere Road.

Map reference:-	J33:534242 (Metric); S058:630487 (Yard).
Catchment area:-	3.9 km².
Period of record:-	July 1971 - Present.
Mean annual flood:-	25.3 m³/s.
Mean flow:-	0.35 m³/s.

Surveyed reach:-

Cross-sections:-	5 along a 103 m reach.
Manning's n range:-	0.032-0.31
Channel description:-	Bed material ranges from pebbles to large cobbles. Banks are thickly bushed, with trees overhanging channel.

Bed Surface Material

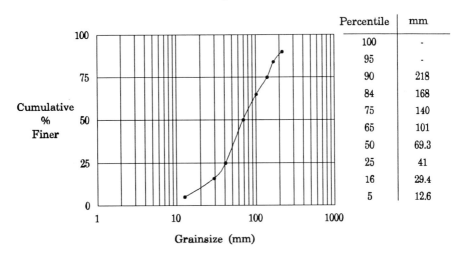

Percentile	mm
100	-
95	-
90	218
84	168
75	140
65	101
50	69.3
25	41
16	29.4
5	12.6

View upstream at top cross-section.

View upstream from bottom cross-section.

n = 0.12

Hydraulic Properties of Reach

Discharge	Water Surface Slope	Friction Slope	Area	Expansion	Hydraulic Radius	Mean Velocity	Manning n	Chezy C	Error
(m³/s)			(m²)	(%)	(m)	(m/s)			(%)
0.02	0.0134	0.0134	0.41	75	0.10	0.07	0.31	2.10	8
0.05	0.0152	0.0152	0.75	-50	0.14	0.09	0.28	2.40	8
0.29	0.0147	0.0146	0.85	-11	0.16	0.36	0.093	7.80	8
1.75	0.0149	0.0146	2.06	-36	0.31	0.88	0.060	13.6	8
1.95	0.0154	0.0150	2.31	-32	0.34	0.87	0.067	12.4	8
1.99	0.0151	0.0147	2.21	-33	0.33	0.95	0.058	14.2	8
4.31	0.0158	0.0147	3.24	-38	0.44	1.38	0.051	17.1	8
4.80	0.0157	0.0144	3.33	-37	0.45	1.49	0.047	18.4	8
12.6*	0.0149	0.0138	6.01	-12	0.67	2.11	0.044	21.5	8
14.5*	0.0145	0.0138	6.29	-3	0.68	2.32	0.040	23.5	8
16.7*	0.0161	0.0135	6.74	-21	0.72	2.53	0.038	24.9	9
18.9	0.0163	0.0121	6.73	-25	0.72	2.90	0.032	29.8	10

* Estimated from rating based on gaugings

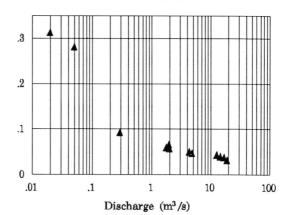

Manning n

Discharge (m³/s)

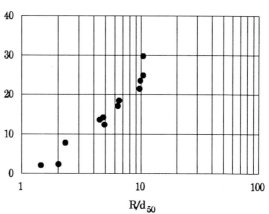

Chezy C

R/d_{50}

292

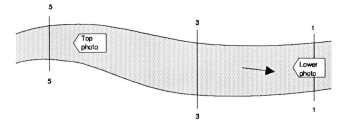

$n = 0.12$

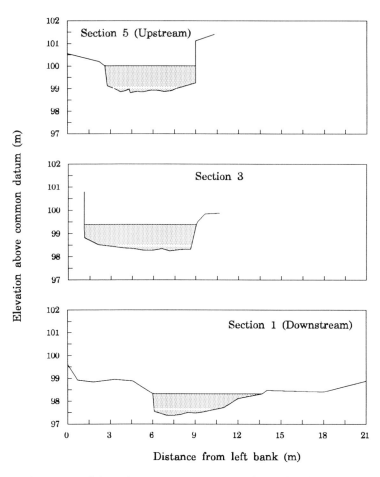

Plan (not to scale) and cross sections, Butchers Creek at Lake Kaniere Road.

n = 0.12

58301: Collins at Drop Structure.

Map reference:-	O27:547052 (Metric); S015:866428 (Yard).
Catchment area:-	17.6 km².
Period of record:-	September 1960 - January 1990.
Mean annual flood:-	30.5 m³/s.
Mean flow:-	0.554 m³/s.

Surveyed reach:-

Cross-sections:-	5 along a 85 m reach.
Manning's n range:-	0.044-0.21
Channel description:-	Bed consists of gravel and cobbles. Banks have overhanging trees, with dense blackberry and scrub on right bank and grass on left bank.

$$n = 0.12$$

View downstream from top cross-section.

View upstream towards top cross-section.

n = 0.12

Hydraulic Properties of Reach

Discharge	Water Surface Slope	Friction Slope	Area	Expansion	Hydraulic Radius	Mean Velocity	Manning n	Chezy C	Error
(m³/s)			(m²)	(%)	(m)	(m/s)			(%)
0.07*	0.00921	0.00915	0.67	-33	0.12	0.16	0.15	4.4	9
0.16*	0.00874	0.00875	1.19	117	0.18	0.15	0.21	3.6	8
0.23*	0.00876	0.00878	1.35	100	0.2	0.19	0.18	4.3	8
0.55*	0.00860	0.00866	1.80	100	0.25	0.33	0.11	6.8	8
2.35*	0.00812	0.00830	3.29	62	0.41	0.75	0.071	12.2	8
5.31*	0.00787	0.00822	4.44	46	0.51	1.23	0.047	18.8	8
13.0*	0.00803	0.00834	7.76	26	0.69	1.70	0.044	21.6	9
30.9*	0.00933	0.00919	15.1	9	0.98	2.07	0.047	21.4	9

* Estimated from rating based on gaugings

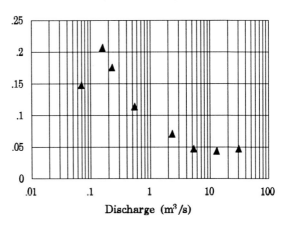

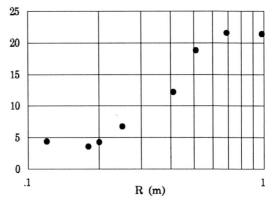

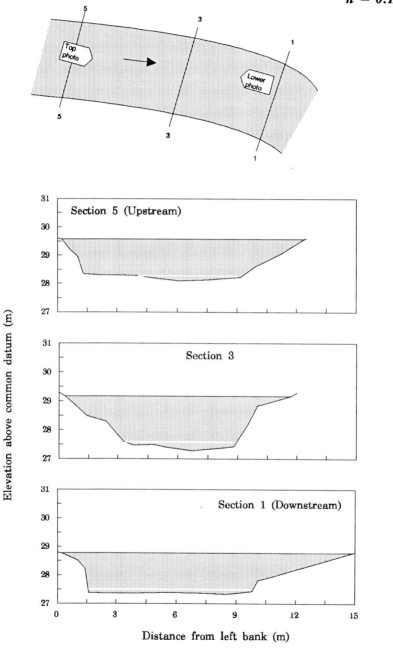

$n = 0.12$

Section 5 (Upstream)

Section 3

Section 1 (Downstream)

Elevation above common datum (m)

Distance from left bank (m)

Plan (not to scale) and cross sections, Collins at Drop Structure.

297

n = 0.13

15453: Waihua at Gorge.

Map reference:-	V16:457208 (Metric); N086:278897 (Yard).
Catchment area:-	44.8 km².
Period of record:-	December 1979 - December 1988.
Mean annual flood:-	42.1 m³/s.
Mean flow:-	1.58 m³/s.

Surveyed reach:-

Cross-sections:-	4 along a 68 m reach.
Manning's n range:-	0.056-0.17
Channel description:-	Bed is composed of boulders which tend to break up the flow into irregular paths. Banks are composed of similar material but are mostly covered by overhanging vegetation and grass.

Bed Surface Material

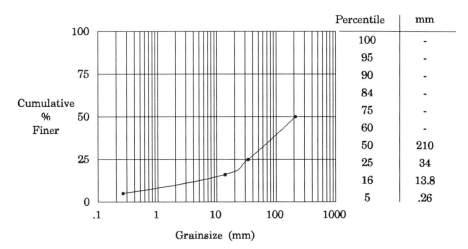

Percentile	mm
100	-
95	-
90	-
84	-
75	-
60	-
50	210
25	34
16	13.8
5	.26

View downstream from top of reach.

View upstream from cross-section 2.

n = 0.13

Hydraulic Properties of Reach

Discharge	Water Surface Slope	Friction Slope	Area	Expansion	Hydraulic Radius	Mean Velocity	Manning n	Chezy C	Error
(m³/s)			(m²)	(%)	(m)	(m/s)			(%)
0.42	0.0124	0.0124	1.94	11	0.21	0.22	0.17	4.5	8
1.10	0.0136	0.0135	2.93	-12	0.27	0.39	0.11	6.9	8
6.55*	0.0158	0.0156	7.42	-18	0.51	0.89	0.087	10.2	8
19.2	0.0186	0.0176	12.7	-28	0.76	1.53	0.068	13.9	8
19.8	0.0152	0.0172	12.8	-26	0.77	1.56	0.066	14.3	8
20.3	0.0182	0.0172	13.2	-25	0.78	1.56	0.067	14.1	8
30.2	0.0195	0.0175	15.3	-30	0.86	2.02	0.056	17.4	8
32.1	0.0201	0.0179	16.2	-32	0.89	2.03	0.058	16.3	8

* Estimated from rating based on gaugings

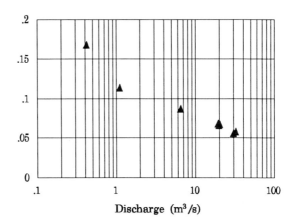

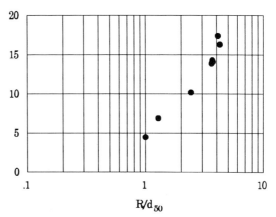

$$n = 0.13$$

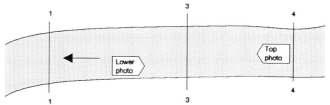

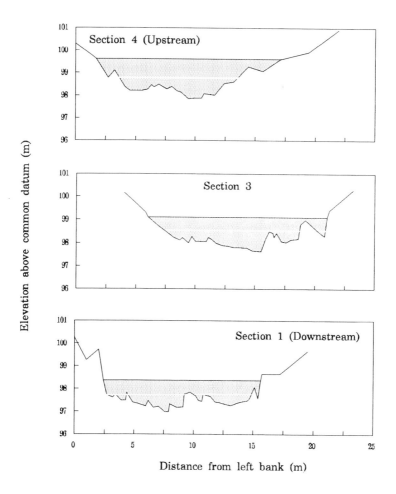

Plan (not to scale) and cross sections, Waihua at Gorge.

n = 0.16

33347: Wanganui at TePorere.

Map reference:-	T19:344368 (Metric); N112:087944 (Yard).
Catchment area:-	28.2 km².
Period of record:-	January 1966 - Present.
Mean annual flood:-	29.8 m³/s.
Mean flow:-	1.3 m³/s.

Surveyed reach:-

Cross-sections:-	2 along a 85 m reach.
Manning's n range:-	0.059-0.18
Channel description:-	Bed consists of cobbles and small boulders. Banks are lined with native flaxes and bushes.

Bed Surface Material

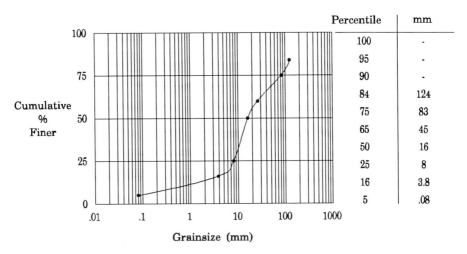

Percentile	mm
100	-
95	-
90	-
84	124
75	83
65	45
50	16
25	8
16	3.8
5	.08

View downstream from middle of reach.

View upstream from middle of reach.

_n = 0.16_

Hydraulic Properties of Reach

Discharge	Water Surface Slope	Friction Slope	Area	Expansion	Hydraulic Radius	Mean Velocity	Manning n	Chezy C	Error
(m³/s)			(m²)	(%)	(m)	(m/s)			(%)
0.93	0.0184	0.0182	3.10	-74	0.40	0.45	0.18	4.7	8
0.98	0.0184	0.0182	3.25	-73	0.41	0.45	0.18	4.6	8
1.17*	0.0189	0.0185	3.40	-74	0.42	0.53	0.16	5.3	8
1.17	0.0188	0.0184	3.45	-72	0.43	0.50	0.17	5.0	8
2.66	0.0189	0.0183	4.50	-64	0.53	0.76	0.12	7.2	8
13.1	0.0189	0.0176	9.10	-38	0.87	1.53	0.079	12.3	8
15.8*	0.0193	0.0175	9.45	-39	0.89	1.77	0.069	14.1	8
16.2*	0.0197	0.0176	9.35	-40	0.88	1.85	0.066	14.7	8
29.3	0.0203	0.0177	12.8	-33	1.07	2.38	0.059	17.2	8

* Estimated from rating based on gaugings

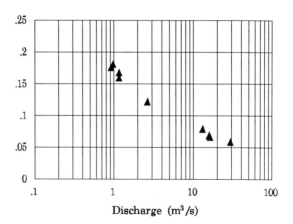

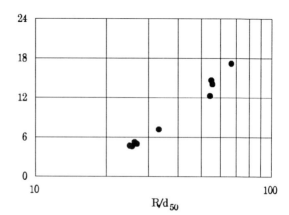

$$n = 0.16$$

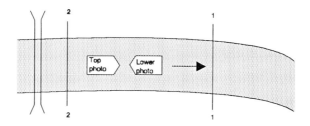

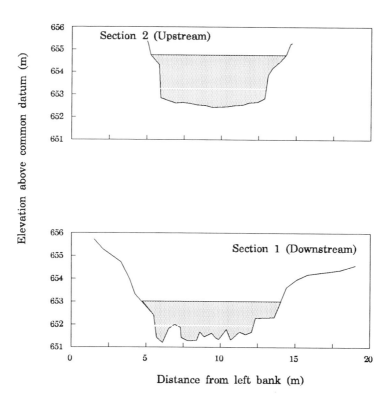

Plan (not to scale) and cross sections, Wanganui at TePorere.

n = *0.18*

47527: Opahi at Pond.

Map reference:-	P05:764436 (Metric); N015:234344 (Yard).
Catchment area:-	10.6 km².
Period of record:-	February 1966 - Present.
Mean annual flood:-	24.4 m³/s.
Mean flow:-	0.27 m³/s.

Surveyed reach:-

Cross-sections:-	3 along a 60 m reach.
Manning's n range:-	0.081-0.23
Channel description:-	Bed consists of sheet lava and silica clays. Right bank has short grass; left bank is lined with sedge and brush.

306

View upstream from middle cross-section.

View downstream from middle cross-section.

n = 0.18

Hydraulic Properties of Reach

Discharge	Water Surface Slope	Friction Slope	Area	Expansion	Hydraulic Radius	Mean Velocity	Manning n	Chezy C	Error
(m3/s)			(m²)	(%)	(m)	(m/s)			(%)
0.25*	0.00052	0.00052	3.50	32	0.57	0.08	0.23	3.9	13
0.31*	0.00033	0.00033	3.79	33	0.60	0.09	0.17	5.4	17
0.38*	0.00032	0.00032	3.95	32	0.62	0.11	0.14	6.5	18
1.03*	0.00038	0.00038	5.37	31	0.71	0.21	0.088	10.8	15
5.80	0.00195	0.00193	9.96	17	1.02	0.61	0.081	12.4	9
5.88*	0.00215	0.00213	9.79	18	1.01	0.63	0.082	12.2	9
7.46	0.00227	0.00222	11.4	13	1.09	0.68	0.081	12.6	9

* Estimated from rating based on gaugings

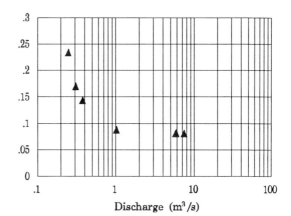

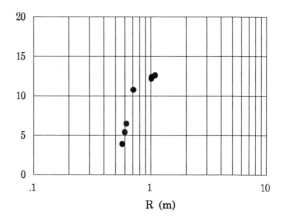

$$n = 0.18$$

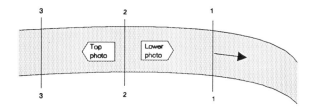

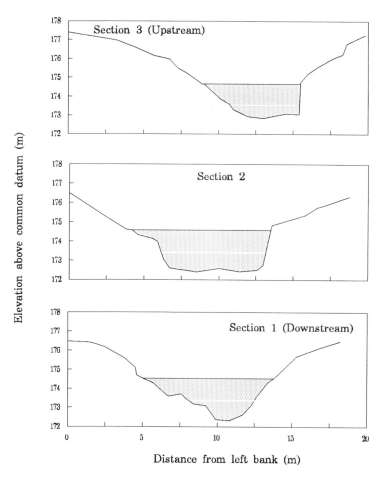

Plan (not to scale) and cross sections, Opahi at Pond.

n = 0.20

46609: Mangere at Kara Weir.

Map reference:-	Q07:226093 (Metric); N020:750983 (Yard).
Catchment area:-	12.3 km².
Period of record:-	April 1975 - Present.
Mean annual flood:-	52.5 m³/s.
Mean flow:-	0.34 m³/s.

Surveyed reach:-

Cross-sections:-	3 along a 100 m reach.
Manning's n range:-	0.041-0.20
Channel description:-	Bed consists of sand and gravel. Both banks are lined with small ferns, grasses, and humus covering a silt loam.

Bed Surface Material

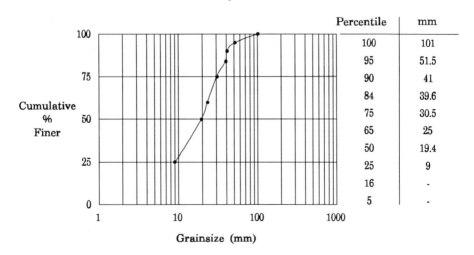

Percentile	mm
100	101
95	51.5
90	41
84	39.6
75	30.5
65	25
50	19.4
25	9
16	-
5	-

Grainsize (mm)

View upstream from middle cross-section.

View downstream from middle cross-section.

n = 0.20

Hydraulic Properties of Reach

Discharge	Water Surface Slope	Friction Slope	Area	Expansion	Hydraulic Radius	Mean Velocity	Manning n	Chezy C	Error
(m³/s)			(m²)	(%)	(m)	(m/s)			(%)
0.65*	0.00389	0.00388	3.52	-14	0.43	0.19	0.20	4.5	8
0.86	0.00403	0.00402	3.78	-15	0.45	0.23	0.17	5.2	8
1.34*	0.00491	0.00488	4.52	-25	0.52	0.30	0.16	5.8	8
6.95	0.00198	0.00206	10.0	35	0.87	0.71	0.060	16.3	8
7.67	0.00241	0.00249	10.6	29	0.90	0.73	0.065	15.1	8
7.69	0.00271	0.00277	10.5	24	0.90	0.74	0.068	14.5	8
9.24*	0.00358	0.00362	13.9	19	1.09	0.67	0.099	10.3	8
12.5*	0.00314	0.00320	17.6	24	1.29	0.72	0.095	11.0	8
20.4§	0.00282	0.00293	21.4	25	1.45	0.96	0.073	14.6	8
29.5§	0.00177	0.00197	25.5	32	1.58	1.17	0.052	20.9	10
46.6§	0.00203	0.00229	33.2	27	1.68	1.42	0.048	22.7	11
87.0§	0.00187	0.00225	48.2	24	1.90	1.82	0.041	27.3	14

* Estimated from rating based on gaugings   § Estimated from theoretical rating for structure

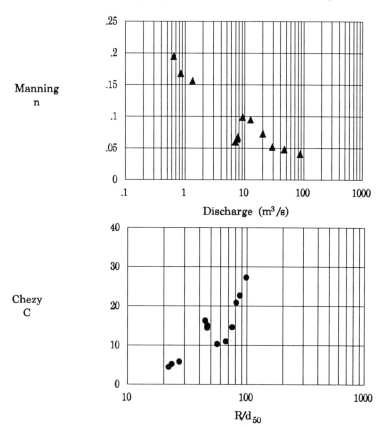

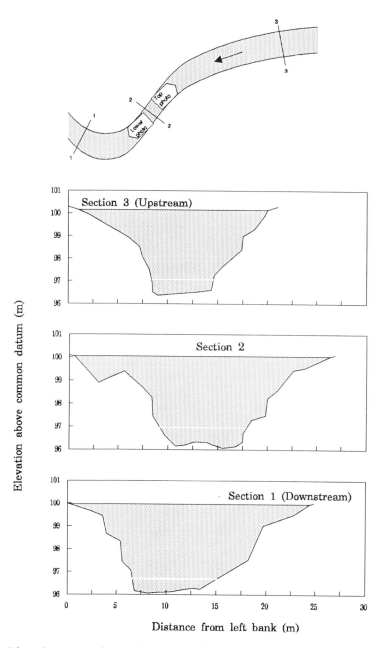

Plan (not to scale) and cross sections, Mangere at Kara Weir.

n = 0.20

39201: Waiwakaiho at SH3.

Map reference:-	P19:082292 (Metric); N109:709822 (Yard).
Catchment area:-	58 km².
Period of record:-	January 1980 - 1989.
Mean annual flood:-	327 m³/s.
Mean flow:-	0.63 m³/s.

Surveyed reach:-

Cross-sections:-	3 along a 100 m reach.
Manning's n range:-	0.047-0.18
Channel description:-	Bed comprises cobbles and boulders, some up to 2m in diameter. Banks are bouldery with occasional scrub.

Bed Surface Material

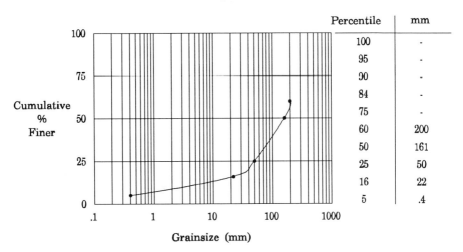

Percentile	mm
100	-
95	-
90	-
84	-
75	-
60	200
50	161
25	50
16	22
5	.4

$n = 0.20$

View upstream at top of reach.

View downstream to bottom of reach.

n = 0.20

Hydraulic Properties of Reach

Discharge	Water Surface Slope	Friction Slope	Area	Expansion	Hydraulic Radius	Mean Velocity	Manning n	Chezy C	Error
(m³/s)			(m²)	(%)	(m)	(m/s)			(%)
2.44*	0.00908	0.00906	8.56	-19	0.45	0.29	0.18	4.8	8
2.80*	0.00903	0.00900	8.98	-17	0.46	0.32	0.17	5.3	8
3.43*	0.00890	0.00887	9.94	-13	0.49	0.35	0.16	5.6	8
9.12*	0.00825	0.00825	17.2	6	0.70	0.53	0.13	7.2	8
21.8*	0.01000	0.00995	24.2	-2	0.92	0.91	0.10	9.8	8
26.4*	0.01050	0.01037	25.7	-4	0.96	1.03	0.092	11.8	8
31.2*	0.01120	0.01100	27.5	-8	1.01	1.14	0.090	11.1	8
77.4*	0.01240	0.01220	40.9	-3	1.30	1.90	0.067	16.6	8
216*	0.01760	0.01480	59.8	-18	1.71	3.64	0.047	23.2	10

* Estimated from rating based on gaugings

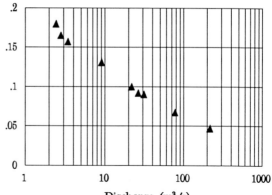

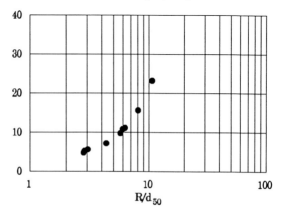

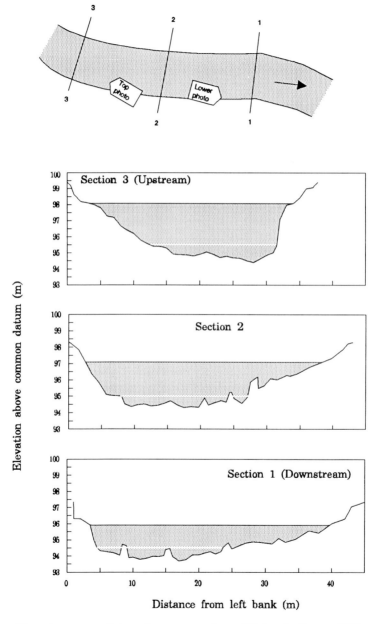

Plan (not to scale) and cross sections, Waiwakaiho at SH3.

n = 0.20

89103: Okarito at Lake Wahapo.

Map reference:-	H35:872692 (Metric); S071:895899 (Yard).
Catchment area:-	51 km².
Period of record:-	November 1967 - Present.
Mean annual flood:-	233 m³/s.
Mean flow:-	10.7 m³/s.

Surveyed reach:-

Cross-sections:-	5 along a 190 m reach.
Manning's n range:-	0.098-0.20
Channel description:-	Bed material comprises large boulders. Both banks are steep, rocky and bush-clad, with trees overhanging the channel. Part of the left bank is a man-made rock wall.

Bed Surface Material

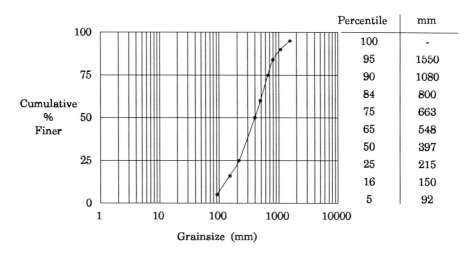

Percentile	mm
100	-
95	1550
90	1080
84	800
75	663
65	548
50	397
25	215
16	150
5	92

View from mid-reach looking upstream towards lake outlet.

View downstream from middle of reach.

_n = 0.20_

Hydraulic Properties of Reach

Discharge	Water Surface Slope	Friction Slope	Area	Expansion	Hydraulic Radius	Mean Velocity	Manning n	Chezy C	Error
(m³/s)			(m²)	(%)	(m)	(m/s)			(%)
17.4*	0.0387	0.0387	22.9	57	0.78	0.80	0.20	4.7	8
59.9§	0.0376	0.0375	39.4	63	1.11	1.61	0.12	8.3	8
154§	0.0369	0.0375	66.4	57	1.59	2.39	0.11	10.0	8
181§	0.0381	0.0385	72.6	37	1.70	2.53	0.11	10.1	8
282§	0.0388	0.0386	90.9	23	2.00	3.14	0.098	11.5	8

* Estimated from rating based on gaugings   § Estimated from theoretical rating for structure

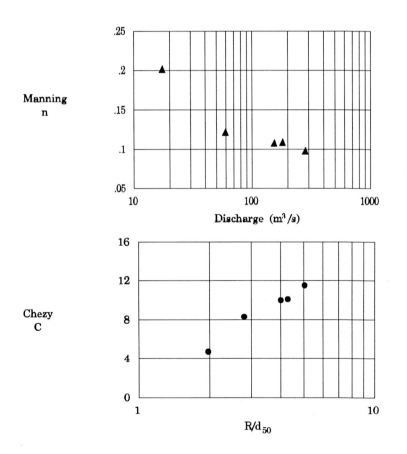

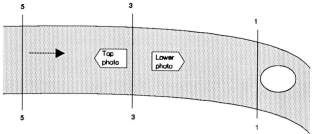

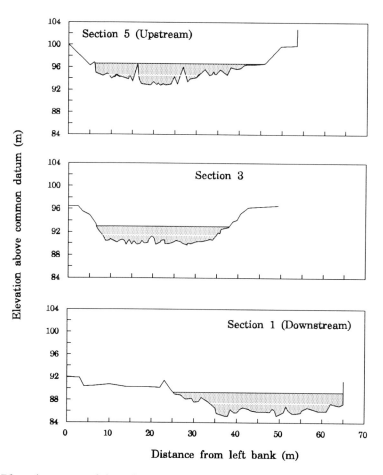

Plan (not to scale) and cross sections, Okarito at Lake Wahapo.

n = *0.27*

67602: Huka Huka at Lathams Bridge.

Map reference:-	N36:937175 (Metric); S094:142293 (Yard).
Catchment area:-	12 km².
Period of record:-	December 1987 - Present.
Mean annual flood:-	12 m³/s.
Mean flow:-	0.17 m³/s.

Surveyed reach:-

Cross-sections:-	3 along a 78 m reach.
Manning's n range:-	0.07-0.29
Channel description:-	A steep gradient stream with large irregularly shaped boulders creating a confused flow pattern of rapids, riffles, and pools. Banks are mainly grassed, with protruding boulders and overhanging trees.

Bed Surface Material

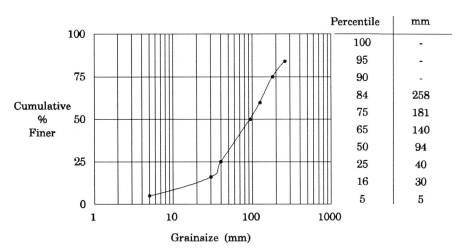

Percentile	mm
100	-
95	-
90	-
84	258
75	181
65	140
50	94
25	40
16	30
5	5

Grainsize (mm)

n = 0.27

View looking downstream at middle cross-section.

View upstream at bottom cross-section.

n = 0.27

Hydraulic Properties of Reach

Discharge	Water Surface Slope	Friction Slope	Area	Expansion	Hydraulic Radius	Mean Velocity	Manning n	Chezy C	Error
(m³/s)			(m²)	(%)	(m)	(m/s)			(%)
0.09*	0.0390	0.0389	0.60	-38	0.11	0.15	0.29	2.4	8
0.48*	0.0406	0.0404	1.62	-39	0.24	0.32	0.22	3.6	8
0.63	0.0406	0.0403	1.76	-38	0.25	0.38	0.19	4.1	8
1.08*	0.0406	0.0402	2.15	-35	0.29	0.53	0.15	5.2	8
1.63	0.0402	0.0397	2.48	-28	0.33	0.69	0.12	6.5	8
1.93	0.0402	0.0395	2.61	-29	0.34	0.77	0.11	7.3	8
3.55*	0.0407	0.0395	3.47	-28	0.42	1.05	0.098	8.7	8
4.17*	0.0418	0.0403	3.67	-30	0.44	1.17	0.092	9.3	8
5.09*	0.0402	0.0389	3.93	-22	0.46	1.31	0.084	10.3	8
8.17	0.0403	0.0382	4.77	-19	0.51	1.73	0.070	12.6	8

* Estimated from rating based on gaugings

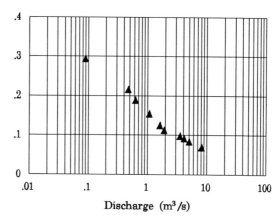

Manning n vs Discharge (m³/s)

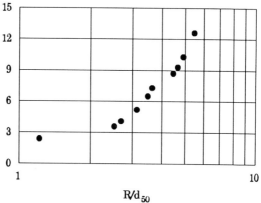

Chezy C vs R/d₅₀

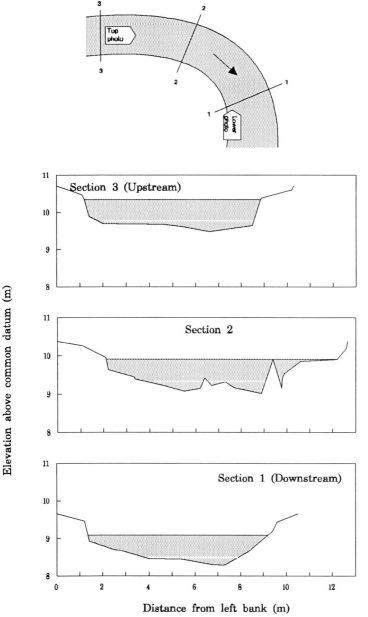

$n = 0.27$

Plan (not to scale) and cross sections, Huka Huka at Lathams
Bridge.

REFERENCES

Arcement, G.J., Jr., and Schneider, V.R. 1989. Guide for selecting Manning's roughness coefficients for natural channels and floodplains. *U.S. Geological Survey Water-Supply Paper* 2339, 38 p.

Arnold, P.E., Holland, G., McKerchar, A.I., and Soutter, W.R. 1988. Water Resources Survey Hydrologist's Field Manual. DSIR Water Sciences Division, Wellington, 148 p.

Barnes, H.H., Jr. 1967. Roughness characteristics of natural channels. *U.S. Geological Survey Water-Supply Paper* 1849, 213 p.

Bray, D.I. 1979. Estimating average velocity in gravel-bed rivers. *Journal of the Hydraulics Division, ASCE,* Vol. 105, No. HY9: 1103-1122.

Chow, V.T. 1959. Open-channel hydraulics. McGraw-Hill Book Co., New York, 680 p.

Church, M.A., McLean, D.G., and Wolcott, J.F. 1987. River bed gravels: sampling and analysis. Chapter 3 in "Sediment transport in gravel bed rivers", C.R. Thorne, J.C. Bathurst, and R.D. Hey (eds.), John Wiley & Sons, Chichester: 43-88.

Cowan, W.L. 1956. Estimating hydraulic roughness coefficients. *Agricultural Engineering,* Vol. 37, No. 7: 473-475.

Griffiths, G.A. 1981. Flow resistance in coarse gravel bed rivers. *Journal of the Hydraulics Division, ASCE,* Vol. 107, No. HY7: 899-918.

Herschy, R.W. 1985. Streamflow measurement. Elsevier Applied Science Publishers, London, 553 p.

Jarrett, R.D. 1984. Hydraulics of high-gradient streams. *Journal of Hydraulic Engineering,* Vol. 110, No. 11: 1519-1539.

Kellerhals, R., and Bray, D.I. 1971. Sampling procedures for coarse fluvial sediments. *Journal of the Hydraulics Division, ASCE,* Vol. 97, No. HY8: 1165-1180.

Keulegan, G.H. 1938. Laws of turbulent flow in open channels. *Journal of Research of the National Bureau of Standards,* Vol. 21, Research Paper 1151: 707-741.

McKerchar, A.I., and Pearson, C.P. 1989. Flood frequency in New Zealand. Publication No. 20 of the Hydrology Centre, Christchurch, 87 p.

Walter, K.M. 1990. Index to hydrological recording sites in New Zealand, 1989. Publication No. 21 of the Hydrology Centre, Christchurch, 181 p.

Water Resources Survey, 1989. Quality manual. Unpublished report of the Water Resources Survey, Department of Scientific and Industrial Research, Wellington, New Zealand, 10 p.

Wolman, M.G. 1954. A method of sampling coarse river bed material. *Transactions, American Geophysical Union,* Vol. 35, No. 6: 951-956.

SITE INDEX

Site name (Site number) .... *Page*

Arnold at Lake Brunner (91405) 138
Avon at Gloucester Street Bridge (66602) 62
Awanui at School Cut (1316) 230
Buller at Woolfs (93208) 146
Butchers Creek at Lake Kaniere Road (90605) 290
Cardrona at Albert-town (75290) 122
Clarence at Jollies (62105) 130
Clutha at Lowburn (75214) 46
Cobb at Trilobite (52916) 258
Collins at Drop Structure (58301) 294
Forks at Balmoral (71129) 186
Fraser at Old Man Range (75259) 242
Gowan at Lake Rotorua (93213) 190
Grey at Dobson (91401) 58
Hakataramea above Main Highway Bridge (71103) 42
Heathcote at Sloan Terrace (666000 - River number) 78
Hoteo at Gubbs (45703) 174
Huka Huka at Lathams Bridge (67602) 322
Hutt at Kaitoke (29808) 126
Hutt at Taita Gorge (29809) 170
Jollie at Mount Cook Station (71135) 198
Kaipara at Waimauku (45311) 234
Kapoaiaia at Lighthouse (37503) 278
Loganburn at Gorge, Downstream (74347) 26
Loganburn at Gorge, Upstream (74347) 134
Mangaheia at Willowbank (18913) 102
Mangere at Kara Weir (46609) 310
Maruia at Falls (93209) 50
Mill Creek at Papanui (30516) 282
Mokau at Totoro Bridge (40708) 182
Monowai below Control Gates (79712) 66
Ngakawau at Lineslip (93901) 274
Ngaruroro at Chesterhope Bridge (23150) 18
Ngaruroro at Kuripapango (23104) 218
Ngongotaha at SH5 Bridge (1014641) 154
Ngunguru at Dugmores Rock (4901) 286
Northbrook at Oxidation Pond (664042) 266
Oakden Canal at Oakden Culvert (68521) 106
Okarito at Lake Wahapo (89103) 318

Site name (Site number) .... Page

Ongarue at Taringamotu (33316) 90
Opahi at Pond (47527) 306
Orere at Bridge (8604) 54
Oruru at Saleyards (1903) 226
Otaki at Pukehinau (31807) 206
Patea at McColls Bridge (34305) 262
Pelorus at Bryants (58902) 270
Piako at Paeroa-Tahuna Bridge (9140) 70
Pokaiwhenau at Puketurua (1043419) 246
Pomahaka at Burkes Ford (75232) 94
Poutu at Ford (1643444) 14
Rangitaiki at Te Teko (15412) 166
Rangitikei at Mangaweka (32702) 142
Rowallanburn at Old Mill (80201) 254
Ruakokapatuna at Iraia (29250) 250
Stanley Brook at Barkers (57014) 210
Tahunatara at Ohakuri Road (1043428) 150
Taieri at MacAtamneys (74315) 86
Taieri below Patearoa Power Station (74319) 82
Tongariro at Turangi (1043459) 194
Waiau at Otoi (21409) 162
Waiau at Sunnyside (79735) 38
Waiau Water Race at Lateral 2 (64616) 110
Waihua at Gorge (15453) 298
Waikato at Ngaruawahia Cableway (43402) 74
Waikohu at No. 1 Bridge (19734) 98
Waioeka at Gorge Cableway (15901) 114
Waipa at Whatawhata (43433) 178
Waipaoa at Kanakanaia Cableway (19716) 34
Waipapa at Forest Ranger (47804) 238
Waipapa at Ngaroma Rd (43435) 202
Waiwakaiho at SH3 (39201) 314
Wanganui at Paetawa (33301) 158
Wanganui at Te Whaiau Canal (33379) 22
Wanganui at TePorere (33347) 302
Wanganui at Wairehu Canal (33359) 30
Whangaehu at Karioi (33107) 118
Whareama at Waiteko (25902) 214
Whirinaki at Galatea (15410) 222

Huka Huka at Lathams Bridge (pages 322-325)
$Q = 5.09 \, m^3/s, \; n = 0.084$